CAMPSITE

LOUISIANA STATE UNIVERSITY PRESS BATON ROUGE

Architectures of

CAMPSITE

Duration and Place

Charlie Hailey

Published by Louisiana State University Press

Manufactured in the United States of America

First printing

DESIGNER: *Amanda McDonald Scallan*

TYPEFACE: *Whitman and Scala Sans*

PRINTER AND BINDER: *Thomson-Shore, Inc.*

Library of Congress Cataloging-in-Publication Data

Hailey, Charlie, 1970–

Campsite : architectures of duration and place / Charlie Hailey.

p. cm.

Includes bibliographical references and index.

ISBN 978-0-8071-3323-1 (cloth : alk. paper) 1. Camp sites, facilities, etc.—Florida. 2. Camp sites, facilities, etc. —Louisiana—Mississippi River Delta. 3. Camping—Florida. 4. Camping—Louisiana—Mississippi River Delta. 5. Florida—Description and travel. 6. Mississippi River Delta (La.) —Description and travel. 7. Hailey, Charlie, 1970– I. Title.

GV198.L3H35 2008

647.942—dc22

2007042883

The paper in this book meets the guidelines for permanence and durability of the Committee on Production Guidelines for Book Longevity of the Council on Library Resources. ∞

For Melanie and for my parents

CONTENTS

ILLUSTRATIONS

ACKNOWLEDGMENTS

Many people have assisted me in this project. William Tilson has provided indispensable critical readings, guidance, and encouragement; and Diana Bitz has inspired me through her vital scholarship and sustained my work with her invaluable mentorship. Ralph Berry has offered consistently productive comments in an ongoing dialogue critical to my development. I have also benefited greatly from ongoing discussions with Nina Hofer and Martin Gundersen, who have motivated me as educators, thinkers, and makers. My other colleagues at the University of Florida have made the School of Architecture an exceptionally rich environment for teaching and research. I want to thank Robert McCarter for his influential scholarship and assistance. Early in the research, Herschel Shepard also set the highest standards and offered many entry points to essential background information. Martha Kohen has provided crucial support and encouragement in the book's postdissertation stages. Chris Silver has also supported the project in its final phases of development.

Through the years of research, many individuals have also contributed to the book, including Judy Tomaini (Rustie Rock), Forrest Bone, Dr. Arthur McA. Miller, Sue Neff, Cindy Russell, and Alex Necochea. Gary Backhaus, Noel Polk, Denise Cummings, Anne Goodwyn Jones, and Jeff Rice have also offered cogent and very useful commentary on the material in chapters 5 and 8. Krystyna Sznurkowski has helped immensely in the project's final stages. I also wish to thank Joseph Powell at the Louisiana State University Press for his intellectual and logistical contributions as the project has emerged into book form. I greatly appreciate the time and effort of LSU's editors and the anonymous readers who have made many important suggestions along the way. Over the years, Jim Adamson and Steve Badanes of Jersey Devil have helped me understand many things about the architectural vocation and its links to context—living with building and camping on site, albeit in a Bambi trailer with an autographed portrait of Divine.

I owe everything to my fellow campers—my wife, Melanie Hobson, and our children, Aidan and Phoebe, who have endured not only planned excursions but also the eccentricities of domestic camping and a sometimes interminably digressive mind. And to my parents for their lifelong love and support.

Chapters 5 and 7, respectively, appeared previously in slightly different form as "Southern Camp(sites): Florida's Vernacular Spaces from John Ruskin to the Tin Can Tourists of the World," *Southern Quarterly* 41, no. 1 (Fall 2003): 75–96; and "At Home on the Midway: Carnival Conventions and Yard Space in Gibsonton, Florida," in *Symbolic Landscapes* (Springer, forthcoming), chapter 5. Used by permission.

PREFACE

Leaving Home
April 27, 2006 (Savannah, Georgia)

I camped with my family during a conference in the spring of 2006. We pitched our tent on site 107, adjacent to the recycling plant on Tybee Island. In the morning, I would speak at the Society of Architectural Historians meeting up the river in Savannah. My topic was the confluence of military camp and university campus.

Looking back on it now, the combination of intellectual pursuit and family camping trip is a suitable relay for the project presented here. In one sense, we camp to escape, to retreat, to "find" ourselves. The camp is then a home-away-from-home where we might rethink a deliberate life. These are the meditative sites of the philosophers' camps in the Adirondacks and the transcendental grounds of Thoreau's crystallized tent. We also camp to find a new collective space where family and society converge. Many of us have been to day camp or grew up attending summer camps, and the legacies of Chautauqua and Oak Bluffs Campground run deep within the American conscience. But the camp is not and cannot be a place of nostalgia. Economics and logistics complicate camping practice in its recreational guises. Camping is not cheap, and we all have difficulty finding time.

As a new strategic ground, the contemporary camp has become a site that we all must consider, not just intellectually, but politically and even spiritually. Camps are local spaces that often demonstrate global situations and forces. At Slab City, those with no home coexist with second-home Canadian vacationers on a decommissioned military camp. The platforms of Manila Village have a smaller socioeconomic range but register a cultural diversity that draws from Philippine, Chinese, and Creole roots. Thirty miles to the north of Manila Village's traces, we are still coming to terms with the Gulf Coast's hurricane devastation and the new forms of urban camping that have resulted.

Camps also encompass the timing of place. Whether at an urban scale on the cusp of city formation or a singular accidental roadside campsite, the process of camping moves from siting to clearing to making to breaking. Spending winters in their Florida camp, Gibsonton's performer residents remake the carnival midway as a seasonal home that is both cyclic and lingering. Also in Florida, the Tin Can Tourists transformed place and time into a home on the road that we still participate in today. We are today immersed in the indefinite, albeit measured, time of the camp. Intentioned or not, we are all camping.

This camping milieu followed the Second World War as wartime exigencies transformed many university campuses into campgrounds of necessity. During the war, American campuses accommodated soldiers, support staff, and students. My paper at the architecture conference sought to understand the significance of this collision of camp and campus. With the influx of returning GIs, university leaders redeployed decommissioned military buildings on university campuses. No longer institutions signifying permanence, postwar campuses were now temporary places of return—thresholds between war and home.

The flexibility of camp permits the layering of places. Camp's grounds are thinly imbricated like the ashen strata of the campfire pit. I did not realize it at the time, but we were camping over Savannah's foundational histories. Our site, on the north edge of River's End Campground, was not far from James Oglethorpe's base camp of 1733 from which he studied appropriate sites for the future city. Before the British camps of the 1750s, it is said that John Wesley held a prayer service at nearby Estill Hammock—his first on the American continent and perhaps the earliest Methodist camp meeting. By 1898, the area had become Camp Graham in response to the Spanish-American War—a use that continued through its recommissioning as Fort Screven and as a military base during World War II.

Temporary Building X on University of Florida campus near University Auditorium and Century Tower, April 20, 1961. University of Florida Archives.

Soon after making camp, we did realize that, although in the shadow of Tybee's historic lighthouse, only a thin screen of scrub oak separated us from the island's recycling center and waste treatment facility. Along with its flexibility and imbrications, the contemporary camp often exists juxtaposed with other, sometimes contradictory, places and programs. Oglethorpe's original base camp became our own field research station for thinking about and experiencing contemporary places. Advertised as the closest RV park to historic Savannah, River's End occupies the space of a new tourism in which beach access, rather than strategic location, mediates place. But before it is rationalized spatially, camping remains first a practice of place-making, however unlikely the place may be. A burgeoning camping public now finds temporary homes along the periphery, in Wal-Mart parking lots as "boondockers" and at the Black Rock Arts Festival as "burners." Some of us still camp to be outdoors, but our attempts occur within a new nature, tainted as much by industrial zones as "no-see-um" insects.

While I was completing this book, I read Colin Turnbull's essay on the camps of the Mbuti pygmies. His work reminded me of the underlying vein of ritual found sometimes unexpectedly in our contemporary practices. Although in most cases the camping milieu has been secularized, I found significant parallels between Turnbull's work on processional ritual and my research, as well as our broader camping lives. It struck me how the anthropologist Turnbull might look to something as ephemeral as the campsite to understand past or distant cultures. Yet it is camping's temporal and physical presence in our own worlds that makes for a ritualized threshold between past and future. Moreover, Turnbull reminds us that, like the Mbuti pygmies, we do not move from here to there when we camp, but we instead proceed from here to here. We never really leave home, and this effectual familiarity works between places as much as it functions across historical times. This movement does in fact involve place. To Turnbull's concept that ritual is a performance that engages the idea of spirit, I would interject place, not necessarily substituting for "spirit" or even invoking the oft-quoted "spirit of place," but perhaps it is place working as contemporary mediator among our many homes.

We had arrived at the Tybee camp, having accidentally passed through the dense urbanism of Savannah. The intervening estuaries had not prepared us for the camping landscape that was our destination—uneven in its topography but more surprising in its postindustrial development. And that is, in retrospect, how we now arrive at the subject of camping—an unexpected confrontation with a not-so-distant past and a tenuously urbanized future. I then understood the camp as the destination of one trip and the onset of another itinerary—one that is equally physical and metaphysical. The camp and its spaces are the quintessential thresholds between past and present and states of the mind and the places we call home. To arrive at the question—how do we make places and homes in these itinerant times?

CAMPSITE

Marion Post Wolcott, "Guest at Sarasota trailer park, Sarasota, Florida, beside her garden made of shells and odds and ends. The camp has a garden club for members organized for the purpose of making the surroundings attractive," January 1941. Nitrate negative. Library of Congress, Prints and Photographs Division, FSA-OWI Collection.

INTRODUCTION

Camping Methods

This is a book about camping. Camping, we reconsider time as duration and we reread place through our own itinerancy. How does camping make new places *through* time? The places of camping are not spatial derivations, but are instead generative of space. Camping begins with place. This emphasis on place-making is not strictly conceptual but arises out of the inherently paradoxical nature of the practice. Oftentimes within a public domesticity, camping works between mobility and fixity, locality and foreignness, and temporality and permanence. At the root of this latter pairing are the cycles, loops, and episodes of camping time. Each camp is chronically becoming. As campers, readers, and researchers, we negotiate these paradoxes of time and place to understand home amidst contemporary itinerancies. Camping is both locative practice and timeless process. This might sound restorative or even nostalgic, but at its core is an interrogation that disallows wistful reduction. Camping practices are complicated by the inherent deracination and the undeniable self-consciousness of the activity itself. Within this framework, we look to camping for deeper ways of thinking about and making place. How can we then understand camping as a method of place-making and of timing place? Such a question asks us to reconsider not only our assumptions about place, time, and home but also method itself.

Camping methods have their origins in the rigid flexibility of the process. We leave home, we arrive at a site, we clear an area, we make and then finally break camp before departing. The phases of this sequence are at once rigorous and indistinct. To understand this paradox, I emphasize the role of process throughout the text. It is possible that this inclination arises out of my work as an educator in architecture—a discipline in which design process necessarily unfolds and builds on itself. But I feel sure that the line of influence runs more deeply in the other direction—camping practices remind us that our places, our built environments, our homes are being made, constructed, and always becoming. The resolute incompleteness of camping parallels our immersion in process. This introduction draws on the tension between leaving home and arriving at a new, oftentimes supplemental, home—a dynamic relation revisited in the conclusion, which will begin again through an inevitable departure, a "casting off." The process of camping practice—siting, clearing, making, and breaking—serves as this work's organizational matrix.

Situating a Camping Practice

Between arrival and departure are the activities of camping. The structure of the camping process allows for explication and invention. As a rhetorical and creative device, camping practice follows a sequence that in part is linear but as a whole forms a cyclic construction in which arriving and departing overlap.[1] Camping proceeds from siting to clearing to making and finally to breaking before an eventual resiting. These phases in the sequence often overlap and are susceptible to interruption, accident, and stoppage. Thus, while delimiting for clarity, the definitions that follow characterize the essence of each term but at the same time refer to their presentation as verbal nouns, the appended "-ing." The idea of action inherent in these verbal nouns is found in their Latinate grammatical term "gerund," which is "capable of being construed as a noun" but retains the "regimen of the verb."[2] The Latin *gerundum* is literally "a carrying on," and it is this suggestion of continuation, of unending process, that is present in the procedure of camping. In this open system, siting continues *through* clearing, making, and breaking. Clearing does not cease with the initiation of the "making" phase and so on. In its implied continuity, the "ing" suffix relates to time, particularly a time of duration in which camp is actually a "camping." These organizing gerunds paradoxically combine the active and the passive, the moving and the fixed, the verb and the noun.

Siting

Yet site does not situate.
—EDWARD CASEY (1997)

Siting is the process that leads up to the establishment of a site.[3] This activity works between the locative, specific, and reified qualities of site and the more open-ended practices of exploration and discovery. Consequently, siting is a negotiation. Methods of negotiation require openness, and siting begins by considering relations between site-driven constraints and external needs: "[b]eing open is setting out the 'facts,' not only of a situation but of a problem. Making visible things that would otherwise remain hidden."[4] In camping, siting entails a decision-making activity. Often presented with a quite literally "open" field, the camper chooses a site in a highly conditional and contingent exchange. In many cases, siting camp depends on the attributes of ground. In camping, qualities of contour, solidity, texture, hardness, and other particularities of the ground cannot be leveled, compacted, or otherwise altered as in typical building projects and sites. Siting instead negotiates the ground.

In camping, this negotiation works as much qualitatively as formally. Under the premise that "site does not situate," place rather than site is what situates the rich multiplicity of things and events. Site remains bound up by and within spatially articulated cartographic and geometric conceptions. Siting is an alternative to the limitations of reductive cartographic, specifically Cartesian, aspects of site. In this context, the procedure of siting camp can be compared to the *situation construite*. For Guy Debord and the Situationists, the construction of a situation was a way of siting that moved beyond the formal and visible characteristics of a particular place. Siting becomes situating. In his initial "Report on the Construction of Situations," Debord noted that architecture "must advance by taking emotionally moving situations, rather than emotionally moving forms, as the material it works with. And the experiments conducted with this material will lead to new, as yet unknown forms."[5] The situation precedes the form. The characteristics of setting up, constructing, or siting a situation mirror the initial procedures of camping. According to a follow-up report produced by Situationist International in 1958, the *situation construite* should be provisional and should be experienced, or lived, rather than merely constructed as a work of art and left to be read and interpreted.[6]

The constructed situation was a provisional experience with many characteristics of the camping procedure and environment. As "an integrated ensemble of behavior in time," this situation was at its core a communal endeavor, and its collective "desire" was then sited within a "temporary field of activity." And, like the operational emphasis in the making of camp, the constructed situation was "composed of actions contained in a transitory décor."[7] Situationism lingers in a chronic siting. The populist precedents of camping were adapted to the avant-garde's own experimentation with place-making. Prewar legislation of increased vacation time in Europe and postwar gypsy camp settlements inspired the Situationists to traverse the emerging psychogeographic contours of everyday life. Camping permutations ranged from the leisure-class British holiday camps of William Butlin to marginalized gypsy settlements in Alba, Italy, where Constant Nieuwenhuys observed the perpetual siting and resiting of the gypsy camp.

The disconnect between siting and site highlights the problem of place in treatments of space and position. Subsumed into sites as nodal positions and with its spaces defined by relativized networks, place becomes directionality and orientation, and its potential to generate meaning is limited.[8] As opposed to this formulation of site, siting retains its active properties. Such openness and nonstatic qualities contrast with the relativism and delimitation of sites found in "striated spaces" and their definition through the "relative global."[9] In the delimited context, sites programmatically function as prisons and penitentiaries, their institutional imperatives leveling the site's ground. In contrast, siting continues its negotiation.[10] Networks of sites rely on their relative position, rather than their locative meaning, for their definition of place. If our time is "one in which space takes for us the form of relations among sites," then the juxtaposition of global sites can yield a placeless heterotopia where the contradictory occupations of site result in conflict and dis-placement.[11] In camping, siting is tied to place-making while site relies on spatial definition.[12]

But if we experience space through site and if heterotopic countersites are closely linked to real places,[13] then negotiating site through place can serve as the basis for determining space. Michel Foucault touches on this engagement in his discussion of inverting the site; but this unquestioned idea of site leaves problematic its role in our actual experience of space and the understanding of place.[14] Accounting for this critique, siting responds to the unresolved role of site and Edward Casey's own claim that site is necessarily antiplace.

Siting occurs simultaneously at two scales: the territorial scale of the camp and the more immediate scale of the body that moves and orients itself within the area of the camp's siting. This dual work of siting responds to the difficulties of resolving the sited stability of Heideggerian habitation with the fragmentation found in postmodern, postcapitalist landscapes of production and philosophy. One alternative to the primacy of space over place is the "non-static anti-site"—a dynamic spatial framework to accommodate fluctuations of place and movement.[15] But the negation of site, whether in countersite, antisite, or nonsite, may not be necessary with the dynamic multiplicity of siting. The inclusion of "site" in this work's title is meant parenthetically and reflects not the complete suppression of site but the possibility of simultaneously maintaining and transforming sitedness through practices of siting, within the overall process of camping. Moreover, the term "camp(site)" also suggests the potential of camp as idea and campsite as an architecture of the incomplete. Such architecture, in its residual siting, waits to be finished, and its incompleteness implies an abundance of possibilities rather than a lack of organization or decisive finitude.

Clearing

Clearing is not simply the removal of obstructions. In camping, places are not cleared away to make room for the fixing of a permanent position. Rather, as a preparation for making camp, clearing is cultivating and gathering. In Heideggerian terms, this process is "thinking the open." Such openness is not attained simply by removing encumbrances but by bringing them to light and by lightening them. In the former, the metaphor of cultivating is associated with tilling that "turns" the previous work of siting and the potential site itself. Clearing, as cultivating, is then a revealing and a disclosing, which can be followed by a drawing together. In its sense of easing a weight, lightening elicits what Heidegger calls the "event of Appropriation"—a pulling together of what is particular and what is near.[16]

Early in my own process of clearing and preparing for this research, I came across the Open City project and its working method, which influenced this work's initial formulation. Although not specifically a camp, the project at the University of Valparaiso in Ritoque, Chile, is a campus with procedural overlays that relate to and frame a camping method. And the Open City exemplifies the lightness of clearing. In the process of designing the campus and teaching design (both closely linked to methods of clearing), ordering principles are malleable and can be transformed as a building of the site proceeds.[17] Idiosyncrasies of contour and ground temper presuppositions of form and design intent. Clearing cultivates the previously constructed situation.

The connotation of *ereignen* as "lighting" expands on the idea of lightening and lightness and develops further the potential for clearing as a conceptual process. In his introduction to *Poetry, Language, Thought,* Heidegger's translator notes the reciprocal nature of clearing as lighting and appropriating. This reciprocity occurs in the "interpenetrating association of coming out into the open, the clearing, the light—or disclosure—with the conjunction and compliancy of mutual appropriation."[18] As place is revealed and opened up, it is also gathered and redefined. But in these events of appropriation, the openness of the cleared site remains. Clearing is "the lighting of self-concealment . . . from which again all self-lighting stems," and this bringing forth "clears the openness of the open."[19]

Such bringing forth and "letting happen" precedes and disallows, or at least postpones, the fixing of place. Such movement is inherent in the activity of clearing. While the clearing of camp may result in the establishment of a bounded situation, the boundary does not delimit or enclose completely. Instead, the boundary, as Greek *peras,* is the place from which "something begins its essential unfolding."[20] Heidegger's *Räumen,* used as "clearing space," thus defines the way that room is made for space. And clearing mediates between the qualities of place, which are gathered, and the required preparations for a space-in-the-making, what Heidegger refers to as *Einräumen.*

The pragmatics of clearing codify the flexible structures of place-making.

Geographic anomalies, political fissures, and aspirations of self-regulation generate camping codes.[21] A process of disclosure, as opposed to restriction, anticipates the making of camps. Strict rules may then be developed, but their origins come out of the cultivation of possibility—an attribute, in the best circumstances, linking camp to place. Other camps derive alternative codes from preexisting urban planning and zoning laws. These revised codes, at the scale of the camp itself rather than its larger territorial context, approach the attributes of the Daoist "way."[22] Between camping and espousing the "way" is a suggestion that clearing is an undoing that has positive rather than negating consequences. The "way" echoes the procedures of cultivating and clearing and relates to a nomadological model: "Thus the martial arts do not adhere to a code, as an affair of the State, but follow ways (*voies*), which are so many paths of the affect; upon these ways, one learns to use them, as if the strength (*puissance*) and cultivation of the affect were the true goals of the assemblage."[23] Code is transformed into "way." Siting and clearing lead to a kind of making defined by the positive application of indirection and avoidance and by instances of direct contact and reference. The "clearing" sections of each case study in this volume outline a forum for the discussion of the codifying influences specific to each place.[24] Similarly, the main "clearing" section (chapter 3) lays out the porous boundary of camp as topic and content. Like the Heideggerian *peras*, this boundary shares the thresholding quality of the camp itself. The activities of defining camp are thus conceived more as starting points than as delimiting lines of inquiry.

Making

The making of camp ties together the broader environmental conditions of each campsite and the more internalized operations that go along with building and occupying camp space.[25] Making is the condition of being enveloped in the process itself, and as a result camp constructions remain in an undeveloped, unfinished, and incomplete state regardless of their apparent degree of permanence. The constructions of *camping* are not things made but are things being made, or more precisely things becoming. Henri Bergson summarizes this activity: "This reality is mobility. There do not exist things made, but only things in the making, not states that remain fixed, but only states in the process of change."[26] In making of this kind, the material forces are emphasized, rather than the effective result or product. The "ambulant procedure and process" of *nomos* contrast with a singular form of *logos* imposed on matter.[27] In the former, materiality is on its way to forming an assemblage, and the latter model submits matter to a series of laws. Given this premise that camping is an immersion in process, what does it mean to build an unfinished architecture, how might the incomplete actually be constructed, and what will it look like?[28]

Although in some usages obsolete, the verbal noun "making" identifies a poetical composition. Camp as method becomes camp as poem.[29] In this historical usage and meaning, the procedure of writing a poem is tied to the activity of making and creating.[30] Also, the resonance between *making* as a poetic act and *poem* as resultant of the making further connects *camp* and poem. At the Open City, the epic poem *Ameirida* serves as the starting point for the actual process of making and building, through the device of the *travesía*. In addition to its connotations in the Open City project, the Spanish word *travesía* refers to a crossroad, crossing, crossway, and journey.[31] Here, the *travesía* serves a dual purpose: to discover valuable connections between the natural and the historical and to inform the ways of making through a process of discovery. The poetic acts of the *travesía* are "group meetings that occur on site and employ poetic methods to initiate discovery and creative processes."[32]

The making of poetry stimulates imagination and initiates construction—*travesías*, as poetic voyages, journeys, and crossings, pull together the immediate as well as the distant. This poetic methodology also links theoretical thought and the pragmatics of concrete action. The concept of *travesía* relates to the way camps are made by engaging "space, place, and poetry through improvisational activity."[33] The making of the Open City also relies on the correspondence between *poiesis* and the understanding of city as *polis*.[34] Like camps in general, the Open City itself then becomes a poetic act. The idea of an "open city" meshes with the understanding of the making of camp not only as an open-ended process but also as an indefinite construction combining material, social, and political forces. In the "open city," the *campo* of the field and country is reintroduced to the *polis*. And making camp returns the *polis* to its historical and poetic origins.[35]

Breaking

Breaking camp returns the camper to pure movement. Tied up in this renewed itinerancy of departure is an assumed arrival. Thus, breaking, in what might be called its "unsiting," retains elements of resiting. Analytically, breaking allows for the possibility of invention. Breaking the ties with a site is a "casting off" that occurs within the space of simultaneous arrival and departure.[36] This exodical space of breaking combines continually localized arriving with the expansiveness of concurrent departure—activities amplified by the multiple campers throughout the network of communal campsites. The acknowledgment of breaking is always present in the creation of "camp space"—an assumed desertion found in Heidegger's ideas about space.[37] To Heidegger's ideas of *Räumen* (clearing space), *Einräumen* (making room), and *Raumgeben* (giving space) is added his notion of *Einbruch*, which is breaking "into" space.[38]

The clearing and making of camp make room for space, and breaking camp makes possible multivalent interactions within a space that is itself being cleared. If clearing makes room for camp in its broader environmental context, then breaking internalizes this process within the campsite's loosely held boundaries. Not only working across scales from environment to specific site, breaking makes "room for leeway" in the definition and use of space. This spatial latitude allows for "diverse engagements" between occupants (as campers), for layered and hybrid programs on a singular ground, and for a closer connection to place.[39] Breaking in effect relinks and refastens camp to place. We might understand this refastening as a process that works back from place to camp. In this sense, camp reifies the abstractions and ideas of place through the camper's mental and often physical reconstruction of place. Through siting and breaking, camp becomes one of many possible materializations of place.

The book's chapters follow closely these phases of the camping process. And in each chapter, this recurring sequence provides a foundation to comprehend how camping repeats itself within the differences of particular places. This consistent structure serves three purposes: to understand the cyclic nature of camping, to explicate how the particularities of the places and sites are construed, and to provide a basis for making connections among the diverse set of case studies.

Three parts further order the book's argument and the sequence of camping. Part 1, "Thinking," includes chapters 1 through 3 and moves from arriving to clearing. In this first part, I propose that camping is a form of "thinking the open." Chapters 4 through 10 form part 2, "Making," in which I argue with each case study that the camp as an intensively made place qualifies the initial openness of the campsite. These chapters are also internally organized by the siting, clearing, making, breaking sequence. Thus, added to qualities of continuation and paradox is the idea that within each phase of camping is found the repeated and embedded manifold of operations. This concept is particularly evident in the making of camp because initial sitings and clearings must be consistently reassessed, and future breakings must be accounted for. In part 3 "Rethinking," I follow up on part 2's fieldwork and conclude with a look at how camping can be understood as a meditative and reflective practice of place-making.

I argue that camps are thresholds. They operate between home and someplace else, and conceptually they work across rituals of making place. To explore camp as threshold and to facilitate transitions, a dated entry precedes each of the book's three parts. Like stories told around the fire, these interludes narrate my own camping excursions. With each of my camping trips, I found unavoidable intersections of thought and practice, thinking and making. The preface having initiated this narrative structure, I hope to camp out with the reader in subsequent interludes, as the book's camping process is thought, made, and rethought in place and time. I begin with camping as thresholding in chapter 1.

To arrive at the practice of camping, chapter 1 explores the multiplicity of architectural thresholds in campsites. In camping, the simultaneity of arriving and departing results in an activity of "thresholding" where time and place often coincide. To explain these ideas, I introduce two analogs: the Airstream Bambi trailer and the navigational stick chart from the Marshall Islands. I argue that the Airstream trailer offers a spatial analog for thresholding while the *rebbelith* charts serve as the operational counterpart. As a practice of thresholding, camping involves thinking generally about how places might be constructed with disparate parts (materials and memories) and specifically about the problem of siting a cohesive difference. "Camp thought" effectively indexes the transformative implications of moving from campsite to campsite.

Chapter 2 locates the project within the organizational and operational context of the camping guide, the manual, and the scrapbook. "Siting Camp" frames the methodological and procedural attributes found within the guide's didactic explication and the manual's grammar of making. Camping relies on both the guide's narration of specific places and the manual's procedural proposals for remaking and inhabiting these places. The structured yet open-ended format of the scrapbook completes the attributes of camp as method. I was not surprised when my research led me to a series of monumental scrapbooks composed by members of the camping group the Tin Can Tourists—remarkable in its intensively collected documentation of campsites and its Ruskinian blending of discursive and digressive narration.

In chapter 3, I provide a working definition of camp. This field of meaning yields a conditional ground that follows up on the situational and operational aspects of camp explored in chapter 2. The working definition links disparate incarnations of camp from the field of the Campus Martius to camp as idea found in work such as Susan Sontag's "Notes on Camp." If the clearing of camp opens up and lightens place, then camping becomes a practice of Nietzschean grounding without ground. To investigate this framework for making places, I provide three narrative threads that discuss how camps form cities: clearing for play, clearing for housing, and clearing an avant-garde urbanism.

Chapter 4 serves as a preamble to the case studies in subsequent chapters, placing each campsite within a temporal context. I present the case that making camp relies on movement from camp to camp and on archaeological times of duration. For comparative purposes, I also introduce the analytic tools used to discuss each camping case study. Derived from how particular camps have been made, I invoke each term to question further the relevance and stability of the made thing. The analytic-poetic terms become "living formatives" that qualify the making of each camp.

In chapter 5, I develop a taxonomy of campsites. As the first case study of part 2, this chapter introduces and defines "camp space" through specific permutations in the Florida context. Based on subtle difference rather than a categorization through simple equivalencies, this taxonomy serves as a flexible grammar by which to understand the Florida camps in particular and the broader sampling of campsites in general. In this chapter, I depart from the typical progression of camping to begin with breaking camp—underscoring the openness of the camping process. These camping spaces occur both as overlays sited on preexisting camps and as "nonlimited localities" defined as much by the minor tourist projects of incidental camping as the more institutionalized modes of tourism found in Florida.

Chapter 6 focuses on Florida's municipal camps occupied by the Tin Can Tourists of the World in the first half of the twentieth century. Exploring the relations among urbanism, early municipalities, and their associated camps, I unpack the telling paradoxes of proximity and distance between citizens and camping tourists. This paradoxical condition has implications for understanding the camps' connection to the urban fabric, their internal architectural constructions that at times transcend their planned ephemerality, and their siting at the confluence of tourism and place. I conclude the chapter with a look at how campers construct "moving images of home" within their home-away-from-home context and through the procedures of making camp.

Chapter 7 continues the story of the Tin Can Tourists to understand the effects of the closure of municipal camps. Evicted from their camp in Tampa, a subgroup of the Tin Can Tourists resettled on an unincorporated peninsula thirty miles to the south. The resulting semipermanent, locally codified development of Braden Castle Park represents a eutopic construct that foreshadows modern suburban insularity. How the community sought to identify itself with the crumbling walls of the site's nineteenth-century sugar plantation house illustrates the tensions between utopia and eutopia through a politically charged appropriation of both space and place.

If the camping tourists latch onto the histories of Braden Castle, then chapter 8's carnival-performer camps develop out of zoning interstices in Florida's rapidly growing cities. In this chapter, I focus on the town of Gibsonton, begun as a fishing camp and further developed around Giant's Camp, the nexus of wintering carnival performers. From the premise that camping parasitism implies the confluence of maintenance and transformation, Giant's Camp is read as an agent of both preservation and change for the larger camp community of Gibsonton. This chapter reviews how places can be understood where, and when, paradoxical attributes intersect: public performance with the domesticity of home, private practices with institutional codification such as zoning, and temporal fluidity within delimited grounds.

In chapter 9, I trace the formation of Slab City from military camp to a strategically marginalized place for Canadian snowbirds and squatters. As the Gibsonton community has reinvented the carnival midway as a home-place, those who dwell at southern California's Slab City construct a self-regulated home in the desert. But Slab City pushes further issues of symbolization and simulation. Through its combination of "dirty realism" and hyperreality, "the Slabs" suggest possible futures of camping as both real and virtual practices of space. And in contrast to its military precursor, Slab City summons a much more subtle, labile code that binds the camp's diverse citizenry. I compare Slab City's place-event to the festival of Burning Man to explore the ways that camping is thematized on the Web and in actual places. The collision of these imagined and actual places sets up the next chapter through the example of Camp Katrina, founded by Burning Man participants as a satellite camp in Mississippi.

In chapter 10, I conclude this series of case studies with the connections between postdisaster New Orleans and Manila Village, a nonextant platform community at the mouth of the Mississippi River. Through examples of the sense of creolized place found historically in Manila Village, I argue that the fugitive forms of the Gulf Coast's postdisaster camps, particularly those in New Orleans' identity-charged urbanism, must allow for connections to place. The place-making sensibilities integral to camping practices might inform ways we transition, however temporally, from displacement to the renewed stabilities of home.

With these ideas, I begin concluding remarks in chapter 11. The movement from permanent to temporary, most dramatically illustrated after Hurricane Katrina in 2005, also characterizes the paradoxes of contemporary urban transformations. In addition to urban growth patterns, I address the scale of the house through the architecture of camping found in R. M. Schindler's exemplary exercise of his concrete "tent." I pick up on Schindler's investigations of the exile's home-place with a rereading and rethinking of Thoreau's camping practices. The hesitation we find between breaking camp and finally departing frames chapter 12's objective to find methods for continuing to explore possibilities of home. Just as the American tradition of camping includes Freud gathering wood for the campfire and Thoreau reading Virgil in his tent, perhaps contemporary camping is a reflective practice for making places anew.

PART I

THINKING

Welcome to Gulf Hills
A few rules:

Keep Stove Clean & ref. defrosted.

Keep garbage out of trailer, Causes Bugs.

No **Drunk** Parties, or **loud** Music.

5 Mi. Per hr. in Park.

This trailer rented to two People only.

We're on Septic tanks, No grease Or food in lines, Nothing but toilet Paper & water in Comode.

Anything Need repairs please let us know.

This is an Adult Park any kids Visiting Must be kept on your lot. My ins. dosent Cover them.

We hope you enjoy what We provide for you.

your dep. Will be returned if Trailer is left Clean & No unnessary damage is done.

Have a good day.

Ray & Wilma Warren

We're here for you.

"Welcome to Gulf Hills: A Few Rules," rules governing Gulf Hills Campground, presented to the author in October 1992. Ink on yellow legal pad paper.

From November 10, 1992 (Santa Rosa Beach, Florida) to 2 a.m., March 13, 1993 (Homestead, Florida)

Gulf Hills Campground, four miles west of the New Urbanist enclave Seaside, was our new home. The ironic distance between the camping code, handwritten on a yellow legal-sized pad, and Seaside's architectural code, which allowed Airstream-type trailers in the town's driveways, was not lost on us. The former, a practical outline for living, and the latter, a closed typological formulary. A relatively easy demarcation, but one that is fundamental to the way we think about the place of camping—critical to rethinking a Heideggerian dwelling, made doubly precarious by a contemporary lack of home and its atrophied public spaces. Camping, we dwell in a public place. Camping, the possibility exists that we are then conscious of our collective homelessness.

What started at Seaside as an experiment in living, modeled on Chautauqua and the Gulf Coast's informal holiday camps, became an exercise in economic determinism. What began as camp meeting, made permanent in how it was to engage thoughts about dwelling, became a suburban coastal resort. Thought about dwelling does not require a definitive lack of home, but latent in our own self-imposed camping expeditions is a need to examine how we live. And yet the exile must return. In August of that year, winds from Hurricane Andrew's glancing blow had forced our temporary relocation to the block walls and collective support of the Salty Dog tavern. By camping out, and then having the luxury of an auxiliary dwelling, we thought about the many locations of home. By camping out, first along two margins—the coastal and the newly urbanized, and then in the posthurricane landscape of South Florida—we remembered we were really arriving at the thought of home.

Our trailer yaws in the early morning winds. The Storm of the Century has forced us out of our Airstream Bambi to traverse the hundred yards to the main house. Sheltered then in what was reputed to be South Florida's only basement, we wondered why we had chosen to camp. Our memory returned to Greece, to Santa Rosa Beach, and to Seaside. Removed again from a supplemental home, our sixteen-foot trailer, we remembered that home is not a singular place, but is perhaps instead the many locations between arrival and departure.

Chapter 1

ARRIVING

Airstream and *Mattang*

> [O]nce thresholds are located, the whole ground plan can be deduced; they provide boundaries that indicate the original layout.
>
> —PETER HANDKE, *Across*

The Airstream trailer and the *mattang* maps of the Marshall Islands are camping vehicles.[1] The former acts as a mass-produced device for travel, albeit phantasmagoric in its reflective skin and recently fashionable as an icon of mobility. A handmade tool for educating young navigators, the latter is also a mental construct, a poetic medium for moving from place to place on the open sea. Both vehicles provide methods of making places and constructing an architectural environment.

I lived in a sixteen-foot-long Airstream Bambi trailer at various times in Florida while learning the art of building.[2] Though the trailer itself remained fixed to the same place, the experience yielded an understanding of what I have called "thresholding." This activity does not so much occur between arrival and departure but instead speaks of the potential simultaneity of arriving and departing. Such simultaneous experience is not enclosed or inscribed by boundaries but instead must occur within the zone of the boundaries themselves. Boundaries do not merely enclose but more significantly serve as areas from which a place opens up or unfolds.[3] Thresholding also articulates an activity defined by coincidences of time and place. Often contradictory, these confluences blur distinctions of present and past, internal and external, and foreground and background.[4] This idea of threshold-

Airstream Bambi trailer in Redland, Florida, 1993. Photograph by author.

ing also relates to method, in particular methods invented to negotiate the paradoxical places of camps. Peter Handke writes of how the archaeologist often begins with the location and excavation of thresholds from which the rest of the building layout can be deduced.[5] These lingering edges indicate how the site was occupied.

In camps, a proliferation of thresholds reflects the occupation of the site, and many of these thresholds are not simply boundaries to be crossed but are entire zones to be occupied. These conditions held a particular fascination for prewar architects. Also finding themselves caught in a newly burgeoning culture of motorcar travel and camping, modern architects such as Rudolph Schindler found the spatial essence of their own camping experiences relevant to housing experiments. In his Kings Road house of 1922, Schindler makes concrete this idea of thresholding in which the grounds of a camp are domesticated and articulated as a series of open threshold spaces that blur the distinction of interior and exterior.[6] The pairing of the Airstream Bambi and the *mattang* map also defines thresholding's combination of the detail and the territory. Within this hypothetical coupling, the Bambi becomes the movable threshold in the multiplicity of possible itineraries held in the navigational field of the Marshall charts.

The scale, spatial economy, monocoque construction, and skin are a few of the Bambi's attributes that characterize the idea and procedure of thresholding. The smallest model of Airstream's fleet, the 1964 Airstream Bambi is 16 feet long and 8 feet wide, with interior dimensions of 7'6" x 12'6" x 6'5". Designers of the original Airstream trailers derived their revolutionary monocoque construction from airplane wing and fuselage structures. The resulting trailer was unique in its completely rounded form, its stress-bearing skin, and the structural synthesis of body and chassis. In what is today identified as the "vintage" Airstream trailer, heat-treated load-bearing aluminum sheeting forms two shells, one inside the other. An aluminum-reinforcing framework binds the shells and holds the 2-inch interstitial space, which is insulated with Aerocore fiberglass. The intensive nesting of shells recalls the original usage of the word "monocoque," which identified a cocoon or the shell of a seed. Itself a shell within a shell, the Bambi becomes a second skin, the mobility of which allows for easy relocation.

As a result of its construction and balance, the trailer can be pulled short distances by direct human force; and a person on a bicycle can pull larger Airstream trailers. To illustrate this ease of movement, 1947 advertisements depicted Alfred Latourneau, the French bicycle racer, effortlessly towing a twenty-two-foot Airstream "Liner" behind his bicycle. The image became one of the Airstream Corporation's primary logos and appeared as an iconic plaque on subsequent trailer models. Just as the trailer's weight can be pulled by a single person, the Bambi's interior width can be spanned by extending one's arms, and its interior length measures only twice this width. This intimate space makes it difficult for a person taller than 5 foot 8 to stand comfortably erect in the trailer. The Bambi amplifies the miniaturization of dwelling space found in Airstreams and other early trailer designs that reflect the confines and configurations of aeronautical and nautical vessels.

Taking into account these conditions, the entirety of the Bambi's interior space becomes a threshold. Here, the use of "space" does not designate the fixed volume defined by the aluminum shell as container. Instead, the trailer's threshold space is a dynamic place of movement. Internally, this threshold space is actually a series of thresholds because there is, out of necessity, a continual movement between and across activities. The everyday activities of sleeping, eating, and washing are collapsed so that spaces normally devoted exclusively to these activities are overtaken by zones in which the actions are mixed. Also, the interior trailer space serves as a semipublic threshold connecting private life to the camp's public zone. More correctly, the threshold spaces of each trailer in a camp extend into the camp's space and bleed into adjacent threshold spaces. Similar to the compressing and collapsing of domestic space that occurs internally, the trailer in its context of the camp serves as a foyer or lobby through which private daily activities fold out into the camp. Thresholding encompasses the overlap of bodily movement within the trailer and the blurring of interior-exterior and public-private zones within the camp.

Thresholding characterizes the residual and continued occupation of the "between space" of the trailer's interior. The apparent volume of the trailer is in reality conceived experientially as an extension of the body—a prosthetic that is built *out from* rather than built *on to* the body. Restriction of movement yields an economy of motion, generating a third skin within the two external skins of the aluminum shell. Because the trailer's scale necessitates an efficiency and compartmentalization of component parts, the phenomenon of thresholding transforms the interior space into a closely knit skin accreted to the manufactured surface of the monocoque shell. In this transformation, the lived space of the trailer-threshold translates the technology of monocoque into a chrysalis (*coque*) woven from the inside through occupation.

In another sense, thresholding sets the limits or thresholds that define the differences, in degree, between inside and outside and public and private. These differences are not absolute, for the spaces of the camp typically begin with the semipublic. The scale of the Airstream Bambi in particular and other larger trailers in general requires an extension of the interior's private threshold space out into the more public zones of the campsite. Because its activities define how spaces of the camp are lived, made, and experienced, thresholding ultimately generates what a particular place, as campsite, means and how its experience is understood *through* time.

If the Bambi is the spatial analog of thresholding, then the *mattang* map offers an operational and methodological companion. My research in preparation for a trip to the atolls of the Central Pacific led me to the indigenous navigational constructs of the Marshall Islanders. I did not truly understand the complexity and subtlety of these maps until flying over the

Pacific expanse and then being deposited on the hook-shaped narrowness of an atoll. For Pacific Islanders, navigation is an architecture. The *mattang* navigation maps were both a product of craft and a guide for crafting. Confluences of tool and method.

On the Pacific trip, I carried two books that also contributed to my understanding of the *mattang* and its context. In *Te Kaihau*, I gleaned the aphoristic statement "Seeing is a matter of faith in sight."[7] And within Umberto Eco's *The Island of the Day Before*, I came across the following less pithy thought: "Hence the maps of the Pacific often seemed arabesques of beaches, hints of perimeters, hypotheses of volumes," and later in the same text: "If Roberto had sensed a world now without any center, made up only of perimeters, here he felt himself truly in the most extreme and most lost of peripheries; because, if there was a center, it lay before him, and he was its immobile satellite."[8]

The *mattang* is a mnemonic device through which the navigator "sees." It is said that a navigator schooled in and faithful to the arts of the *mattang* can navigate without visual faculties. The *mattang* holds seagoing conditions and relations between phenomena in its assemblage of sticks and shells. With these factors in mind, the navigator then reads the wave patterns by sensing the wave forces and swells against the side of the hull while lying prone in the canoe. The Marshallese pilot does not lay out a course or use the maps as aids to recognize or visually identify particular land forms but relies on a combination of empirical data and concealed, nonvisual information that must be understood directly.[9] The position of islands and the depth of the water are intuited by way of forces that occur tangentially through the refraction, reflection, diffraction, and dissipation of wave swell energy along or between islands. The reading of these forces does not privilege visual information and places an equal if not greater importance on senses of hearing and touching.

The Pacific expanses prohibit navigation by seeing objects. Because the Marshallese navigators cannot see "that," they must feel "this"—a more proximate reality of sound and touch. The intensity of the wind, the degree to which the sun burns the skin, the closer call of seagulls, the greater pressure of waves on the side of the hull and its higher pitch or more frequent phase. This synesthetic experience redefines a cartography of sound, touch, and sometimes taste and smell, and yields a topologically defined mental map. The intensities of these phenomena make up the mariner's direct experience (its specificity and "thisness") but also give a network of haecceities that facilitate movement from place to place.[10]

For the practitioner of the *mattang* map, the haecceities and their related forces occur along perimeters and within edges. This situation of being lost among edges that Eco's character laments is actually an assumption necessary to the successful functioning of the *mattang.* The condition of habitual displacement does not constitute being "lost at sea." The forces of displacement paradoxically become locative devices. And the only center among this proliferation of peripheries (in a regional territory, the surface of which is 91 percent water) is the navigator himself or herself. The *mattang* map exemplifies speed and the "individuation" of the mariner's, or traveler's, body. With perimeter as site and haecceities as the qualifiers of this site, the navigator refuses to depart and as a result is always arriving within the open spaces of the sea. The constant arrival affords an orientation that disallows being lost and provides localization without delimiting the seascape. The *mattang* equips the navigator with the local operations necessary to carry out the absolute movement that can be found in the concept of "speed."[11] These local operations are directly tied to the qualitative attributes of haecceities, which are received as intensive phenomena rather than being analyzed as relative movements.

The experience of speed is in marked contrast to our own cultural understanding of rapid movement or the relative passage of time. During the trip to the Pacific, adverse maritime conditions caused my wife's family to be "lost" at sea for more than forty-eight hours. Upon their eventual arrival at their destination on the outer island, family members marveled at the islanders' meditative calmness throughout the passage—contrasted with their own fear because of a protracted travel time, more than ten times the typical period. This waiting as a kind of chronic arrival establishes a local absolute within the sea's expanse.[12] When such navigation is carried out in relative time, this seeming hesitation in open space is frightening and vertiginous—as in the case of my wife's family, who read their journey through the recollection of a hastily receding departure rather than thinking through the phenomenal place of an imminent arrival. This localized absolute, linking haecceity and speed, is analogous to the schema of the thresholding campsite, in which the Marshallese feel so much at home.[13]

In essence, the Marshallese pilot travels in place, from the dynamic position of this local absolute. From a navigational perspective, the sequence of forces that arise from island edges and peripheries moves *to* the craft as a series of thresholds. From the mariner's perspective, these threshold conditions revolve centripetally around the seemingly stationary pilot in a microcosmic version of a Ptolemaic system. The movement-in-place from threshold to threshold is also characterized by a fragmentation measured by the space between phenomena, or "sets of relations," associated with the local absolutes. Such relational traveling moves "from campsite to campsite": "On the nomads of the sea, or of the archipelago, José Emperaire writes: 'They do not grasp an itinerary as a whole, but in a fragmentary manner, by juxtaposing in [the] order [of] its successive stages, *from campsite to campsite* in the course of the journey. For each of these stages, they estimate the length of their crossing and the successive changes in direction marking it.'"[14] Each episode or fragmentary piece of phenomenal data provides a "stage" of experience. In this context, the term "stage" is not a static resting place but the procedural zone between "stops" or "standing places," much like the theatrical and filmic stage serves as the locus of action during designated acts or episodes.[15] Just as the stage or station is process for the Pacific mariner, the campsite becomes the procedural and operational ground for the camper. And the activity of camping occurs between campsites, within the collapsed moments of arrival and departure.

As pedagogical models, the *mattang* and its variants present this observable and implied data in their physical composition. The *mattang* is a predominantly symmetrical model illustrating general concepts of wave action. Its component parts are flexible sticks that can be bound at their intersection with sennit—the thin cordage of braided coconut fiber. Within this woven construct, sticks that are completely wrapped in the sennit allude to minor asymmetries that designate particularities of wave refraction. In some cases, a wrapped stick indicates the direction of the dominant trade wind swell called *rilib* (meaning "backbone").[16] While the *mattang* provides generalized nautical conditions, the *meddo* and *rebbelith* stick charts portray specific islands and island chains within the Ralik and Ratak archipelagoes.[17] Small cowry shells lashed to intersections of sticks indicate island locations within the model; these positions do not show true distances and directions but connote positions understood through wave action and ex-

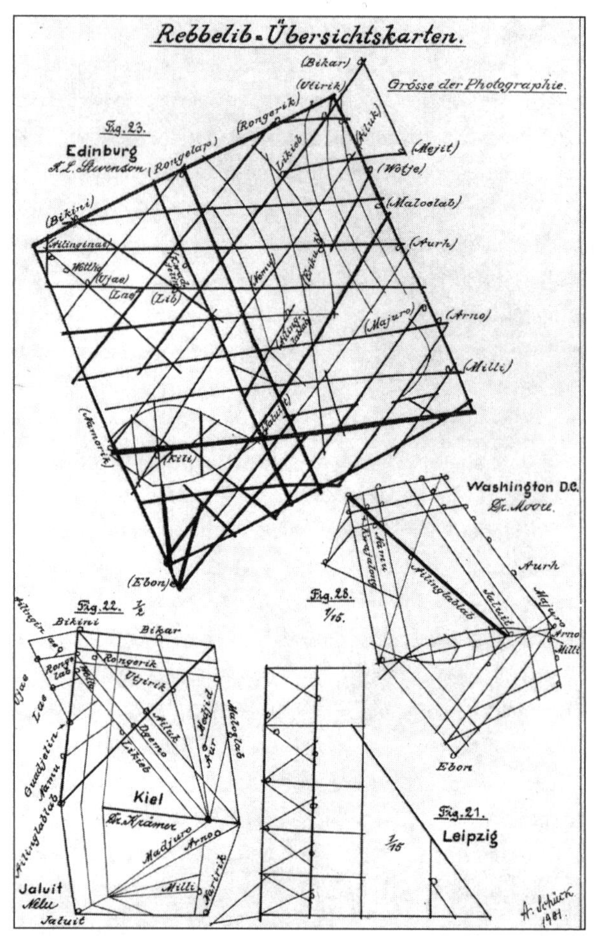

Rebbelith navigation maps with archived location of each artifact, including Berlin, Honolulu, Leipzig, and Hamburg. Albert Schück, *Die Stabkarten der Marshall-Insulaner,* Hamburg: Kommissionsverlag von H. O. Persiehl, 1902.

perienced time. Through a series of "indicator mnemonics" (*rojen kôklôl*), this rendering of perceptions combines with knowledge of wave swells, bird flight patterns, and visible features of distant islands such as trees or atoll rises. The *mattang* indexes a poetic and highly narrative description of the wave pattern components and signs (*kôklôl*). *Rolok* is something lost, *nit in kot* is a hole, *okar* is a root, *bôt* is a knot or node, and *jur in okme* are stakes.

The constructed maps are not used in the actual navigation and are discarded after teaching exercises and memorization.[18] The patterns and relations illustrated by the charts must remain lodged in the oral tradition and cultural memory of the islanders. Respected as spiritual leaders, the teachers maintain this knowledge of the navigational arts as a mnemonics of the sacred. The navigation charts further exemplify the indivisibility of landscape, or more precisely the seascape, and human experience within the Marshall Islanders' culture. The layering of cultural memory creates a foundation for the highly charged, sensual existence that is critical to the Marshallese experience: "Before it can ever be a repose for the senses, landscape is the work of the mind. Its scenery is built as much from strata of memory as from layers of rock."[19] As a mental construct, the landscape mapped by the traditional sea charts combines the physical and the metaphysical—an interlacing evident in the mythically inspired songs of the Marshallese navigator and in the woven lattice of the chart itself.[20] Closely tied to a formulaic system, the songs of the Marshall Island mariners are also navigational reminders and ways of maintaining confidence during the journey. The magical properties associated with the songs and their particularities of arrangement and rhythm affirm the navigator's security and orientation. Songs of navigation string together a limitless array of places, directions, natural forces, and events that measure the journey: *lijiblili ekejeri wa kein, o-o-o-o-o; eato ealok ion; eatoen mij in,* translated as "current whorls which change [the] course of these canoes, o-o-o-o-o; northwest current, that causes death."[21]

The Marshallese navigator moves within this series of thresholds outlined by the *mattang*'s mapped relations and is constantly arriving and departing in response to "hints of perimeters" and "hypotheses of volumes." This centrifugal thresholding, within the open nautical space, complements and contrasts with the more centripetal construction of the Airstream. Also creating a memory theater, the compartmentalization and enfolding scale of the trailer is a physical reliquary in contrast to the mental, or even metaphysical, construction of the *mattang*. Just as the Marshall Islander can lie in his waterborne craft and direct its motion across the open sea with eyes closed, the Airstream dweller can navigate the trailer's confines. In each case, living and the experience of place occur in the interval "from campsite to campsite"—a threshold between going and coming, departing and arriving.[22]

Wally Byam with Airstream Globe Trotter, designated the "World's Most Traveled Trailer." Courtesy of Airstream Corporation.

This activity of "thresholding" yields a typological array of thresholds characterized by the connections to be made between voyager and place, between camper and campsite. The *mattang* connects the navigator with the forces of the sea, the physical with the metaphysical, and the haptic world of touch with that of sight. The trailer links site to place, home to region, and mobile unit to ground. As a result of this connectivity, these details, as *mattang* and Airstream, relate to, circumscribe, and in many ways become territories. The *mattang* becomes the open sea, and the surface of the trailer registers its journey by indexing the camper's itineraries. The "World's Most Traveled Trailer" references past destinations in a listing applied to the exterior skin of the Airstream Globetrotter. The shell of the Airstream trailer does not change and allows for immediate occupation of the site, similar to the rapidly erected archetypal forms of indigenous dwellings such as yurts, tepees, and bamboo lodgings. The context of the trailer's siting does change; and in the particular case of the Airstream trailer, the surface reflects the new location in its polished skin—a mirroring that alludes to the transformative potential, physically and metaphysically, between the new site and its displaced occupant.[23]

Both *mattang* and Airstream establish vectorial conditions. Such vectors, as vehicle and indexed movement, set up connections between particular places. Vectors have the potential to link camp to site and the thinking of places to their making.[24] Although retaining directionality, this type of vector does not link fixed point to fixed point. Instead, the intensification of this directionality becomes indexed travel, recording changes in place and time. As vectorial gestures, the relations between place and time occur at the intersection of thinking and acting; and gestural difference results from the specificity of the nomadic subject's particular transformations, wanderings, and "crisscrossings." Such vectorial thought takes into account external forces through a tactical, as opposed to strategic, mobility that allows for topological, rather than strictly topographical, movement across time and space.[25] "Thinking the vector" allows for the simultaneity of arriving and departing. As mariner, tourist-traveler,[26] and camper, the nomad is a "vector of deterritorialization" transforming places at both a global and a local scale.[27] The camper-traveler thus makes the camping territory through local operations and a globalizing vectorial thought.[28] In its most flexible form, the nomadic camp is not rooted to a particular place. Instead, the multiplicity of camps form an open network, within which the nomadic subject makes connections through repeated but differentiated operations. The camp is the site of nomadic experimentation where thinking and making, arriving and departing meet.

If the vectorial method provides an analog for what might be called "camp thought," it is the rhizomatic exercise that frames an understanding of "camp construction." Camping occurs between the activities of reterritorialization and deterritorialization—operations by which it is defined. And the time of camping is a question of arriving and departing—a becoming rather than a being. If my image of the Airstream is a vector, then the *mattang* is its map. And it is ultimately through this simultaneity of arriving and departing and coincidence of detail and territory that the idea of thresholding resonates with camp. The procedure of camping includes remaking an assemblage and rethinking a place. Camping mixes thinking and making within the campsite, or "place threshold," that simultaneously must identify the characteristic attributes of a specific place and identify *with* the previous experiences of a distant, however similar, place. This mixing, as thresholding, characterizes how things are reassembled for each new campsite and what relations are mentally drawn with previous sitings. In the former, the adjustments to an assemblage or condition are minor, not unlike the navigator's shifting his or her weight to change the vessel's orientation. In the latter, recollection of past camping experience might invoke such reverie as "I remember that last time we camped next to the exposed limestone heated by the afternoon sun." Thresholding and camping require this negotiation of making and thinking. And if camps index both movement and transformation in their sedentary moments, then this indexed movement proceeds both from campsite to campsite, in the series of interlaced sitings and places, and from camp(site) to campsite, when mobility is tempered by a degree of permanence.[29]

From Camp(site) to Campsite

> At the sea.—I wouldn't build a house for myself (and it is part of my good fortune not to be a home-owner!). But if I had to, I would, like some Romans, build it right into the sea—I certainly would like to share a few secrets with this beautiful monster.
>
> —FRIEDRICH NIETZSCHE, *The Gay Science*

What happens when camps become camp-sites, or when the impermanent attains a degree of permanency? In an aphorism typifying his *gaya scienza*, Nietzsche is drawn to the "beautiful monster" of the sea as a home unburdened by rites of ownership. And an early American autocamper will confuse home with the camp of her family's vacation. Her journal reads: "Right after supper we talked a while then went home (I mean back to 'Desota Park')."[30] Nietzsche at home on the sea, and Ruth Deering at home on the road. Can such mobilities of home be retained in the grounds of semipermanent camps or with home ownership? In many cases, the sequence of camping grounds its participants *between* homes. The Airstream trailer was advertised as the "home away from home," with the assumption that the trailer owner could and would always return to a primary home.[31] What this promotional catchphrase does not account for is the coincidence of homes, when primary home and secondary homes become Home. Camp resides within this confluence.

Another point of this inquiry's origin, one much less remote than the Pacific travel, was my purchase of a house as a renovation project and subsequent thoughts about its implications for home. To live in one's work mentally, figuratively, and physically allows the possible confluence of distance and proximity through the work's imagined and actually constructed ideas. The overlap between the renovation project that was the house on Livingston Street in Madison, Florida, and the architectural construct, which this written project has become, lies in each project's negotiation of theory and practice, of imagination and reason, and of mobility and stability. I am reminded of the German Situationist Günther Feuerstein, whose own apartment became the experimental site not only for framing artistic and theoretical expression but also for living. These intensely personal projects, termed "impractical flats," map the artist's ideas, dreams, and history in a domestic palimpsest of the unfinished. Inhabiting our unfinished house, we did not share Feuerstein's outright rejection of labor-saving devices and contempt for environmental comforts, but we did see this domestic occupation of incomplete space as a chance to construct our own real and imagined homes.[32] Just as this research served as the ground for exploring the notion that place can be constructed from disparate, at times placeless or dis-placed, components and ideas, our house became a construction site for nomads at home. We were "building the site" in which home might converge with the permanence of the unfinished house and the temporality of our propensity to drift.[33] We were in effect camping at home.

The idea of camp and camping remained in the unfinished walls and the living spaces, completed only by temporary porches awaiting permanent roofing.[34] If home can be seen as a rhetorical territory, then camp and its image serve as deterritorializing influences that maintain an unfinished ordering in the potential, however latent, transformation of home and ideas about home.[35] With our entry into home ownership on August 28, 1999, the "grounding" of a house's spaces provided both antithesis to a previously peripatetic existence and stable grounds within which to work on our respective dissertations. Such antithetical and paradoxical territories are based on the relational rather than the strictly formal. From a social standpoint, situations in which both communal and economic guidelines are required and mandated through an overlaid set of rules often force a devolution into a solidarity of the same; and the camp becomes an encampment. But the opportunity exists that the camp is a place of inclusion rather than exclusivity and a space of cohesive difference rather than enclosed

Demonstration of lightness at Airstream manufacturing facility.
Courtesy of Airstream Corporation.

consistency. In terms of place and time, the privileging of the relational opens up the possibility of a domestic architecture of the unfinished. Such designation as "unfinished" does not provide valuation or quantification of completeness or of a lack. The unfinished is not necessarily the incomplete. And, at this arrival and departure, the question remains, as it did for Nietzsche, can there be grounding without ground?[36] Can there be a home on the metaphorical sea?

Place, Time, and Vernacular

Camps and campsites occur within the triadic relation of place, time, and the vernacular in the architectures of home. The mobile, yet place-specific, nature of camping underscores the inadequacy of studying temporal built environments by privileging the role of space. The making and remaking of the architectures of camp arise not from the modernist mantra "space, time, and architecture" but from the subtle and flexible triad of place, time, and vernacular.[37] The siting of camp differentiates place from space and at the same time provides a ground, a critical and physical place, for the generation of space. Only in power-driven structures of encampment does space subsume place. Most camps reverse this trajectory such that "the ultimate source of spatial self-proliferation is not the body or the way the world is but the placialization of space itself."[38]

Within this fundamental distinction between space and place, camps establish the conditions for events to occur. Arriving at camp precedes both place-making and events associated with occupying the place. Given this necessary chronological sequence, the dialogue between place, camp, and event is multidirectional. Place results from the "production of an event" that includes the camp and its related minor events—whether setting up a tent, telling stories around the campfire, or waking up to a downpour of rain.[39] The intervention of the camp also reconditions the place, certainly generating a kind of spatial occupation but also adding to the site's layered history, which is densely stratified if it has been intensively occupied. Though its disruption may be lightly drawn and is often ephemeral, the camp is an active protagonist in the immediate environment. The camp construction is a "place that confronts spaces and actions" within a larger, more extensive "landscape of events."[40] Camp's relation to the event not only further differentiates space and place but confirms the significance of camping's temporal attributes. If camping occurs within a landscape of potential events, then the phenomenal contours of place anchor the construction of the campsite in a time of duration. When space is recast as a temporal phenomenon, duration becomes a confluence of simultaneities such that "the only relief" in physical and historical landscapes is "that of the event."[41] Ultimately, the possible confluences of temporality and place pose a problem for the construction of built environments—a complexity that camps already address out of necessity.

Problems occur in reading the architecture of camps when placelessness is assumed under the conceptual mantle of "ephemeral architecture." Recent discussions of ephemeral architecture discount the possibilities of place in the siting and making of impermanent architectural constructions.[42] As a part of this misconception, the often-unexamined attachments of the ephemeral construct to the particular place include not only material connections but also historical, phenomenological, and metaphysical associations. Demountability does not deny the possibility of these linkages, and attachments of a particular duration, however tenuous, amplify the connectivity to a place. For example, the fabric draped on, between, and within the early automobiles and their campsite's context becomes a binding agent between campground and place. Clotheslines, wire fences, branches, fenders, trunks, gravel drives, tree shadows all combine in a woven construction of the site—one from which disengagement appears difficult, if not hesitant or doubtful. Moreover, ephemerality in effect transcends temporality through duration. Stretched in tension and molded to natural and man-made surfaces alike, the ephemeral fabric holds together a very particular, though provisional, assembled moment of place.

In other cases, an intended ephemerality gains a degree of permanence and becomes a site for continued architectural production. In communal camping assemblages, a complex network of related social, cultural, and political productions accompanies this semipermanent siting. Proximity of the mobile and the fixed results in a "vagabond architecture"—an apparent oxymoron in which ironies are redoubled.[43] The vagabondage of a camp's ephemerality is not completely itinerant, and the static nature of architecture is not inflexible. The spaces of camps accommodate the latent freedom and idleness of both vacationer and vagabond. Whether short-lived or having a longer duration, campers occupy an intermediary time, understood in this context as the absence of something else. The intercalary implications of *vacare* reframe the problematic confluence of camp's mobility and fixity in terms of time. The heterogeneity of the camping space, illustrated in the unlikely accommodation of both vagabond and vacationer, defines a fluidity of time as duration.[44] Just as Henri Bergson called for a "new type of space" altered by time, camps arise from the idiosyncrasies of a particular place and the temporal experience of that intermissive site.[45]

The duration of camps, as neither exclusively mobile nor fixed, redefines vernacular architecture as a process. Such a recalibration to understand the vernacular shifts emphasis from building form and its programming of space to a more inclusive treatment of the complexities of context and place.[46] With an inductive approach, the vernacular is both a visual and oral construction.[47] As "vernacular architecture," the visual approach yields an open extensivity as long as regionalist tendencies to replicate the "local architecture" are not fixed in a project of restorative stability. As "architectural vernacular," the oral entails a flexible grammar that, in its inflections, gives an intensivity from which localized methods of experiencing and understanding place might be derived. To avoid the predicament of "new vernacularism,"[48] in which motifs and forms are appropriated as symbolic of place-making, the dialogue works between the large and the small in scalar shifts and between extensivity and intensivity in rethinking subjectivity and place.

The vernacular becomes a dialogue between detail and territory. In architecture, the concept of the tectonic works within this kind of dialogic framework. With camps, tectonic presence is studied alongside the "experience of absence" that characterizes the poetic contours of everyday life.[49] In this formulation, the vernacular is seen as a process that, although in many ways absent, traces or indicates *how* a construction has been made. And buildings and constructions are circumscribed by scalar shifts in an attempt to understand the locality and globality that characterizes contemporary vernacular production. John Brinckerhoff Jackson summarizes this problem of the vernacular by noting the paradox that many of the materials and techniques assumed to be locally derived and indigenously crafted are actually imported "from elsewhere."[50]

The problem of defining home follows closely along the lines of this paradox of a locality constructed of and defined by the distant and the foreign. Notions of home, not necessarily connected to geographies of place, require alternative grounds for speculation. For bell hooks, this home-place is "that place which enables and promotes ever-changing perspectives, a place where one discovers new ways of seeing reality, frontiers of difference."[51] Thus, home still resonates with place, but through modalities and materialities that cannot always be assigned to locative constructions or philosophies.[52] hooks's connection of home with discovery proposes a productively unstable home, the grounding of which occurs through unearthing and maintaining variation.[53] A similar dialogue between camp and the revised home-place suggests ways of defining the paradoxical home, itself cast within a semipermanent milieu.[54] hooks's invocation of "discovery" yields an important general grounding for this work—its theorization of camp is not merely critical but pre-positional, exploring what leads up to camping form and how we arrive at camp.

Amidst contemporary problems of place and time, camp offers an inherently flexible format and generative procedure. Housed within an adapted camping manual, this explication of the camping practice addresses how places have been made and how future place-making endeavors might be framed. As a productive interdisciplinary hinge,[55] the idea of camp is also enlisted as a procedural heuristic that allows for both the discursive and the digressive with the common purpose of developing a method to work between these characteristics. Consequently, the project's originary concepts and their cross-pollinating experimental ground may be sited in what has been called the "peripheral middle," itself held within a "patchwork mentality."[56] Camp as phenomenon, process, and mentality. This connotative field circumscribes the investigation of how movements from camp(site) to campsite redefine place and duration. Camp as subject matter and camp as method.

Chapter 2

SITING CAMP

Guide-Manual-Scrapbook

There is no history book—just a scrapbook of cherished fragments.
—BRITISH BROADCASTING COMPANY (1933)

After arriving at a particular place, siting camp begins. Having arrived at the premise of camp as method, I have found the links of camping form with content and the connections of campsite's interpretation with its experience to be inextricable and to warrant a suitable format. I have sited this treatment of camp within the sequence of camping practice and within the structure of the camping guide and manual. This twofold siting can be compared to backyard camping. When we camp with the family on the lawn, we are carrying out the camping sequence not as simulation, because we really are sleeping outdoors, but as a temporarily defamiliarized space within the recognizable environment of home. Camping at home, we might practice setting up a newly purchased tent, prepare for more far-flung expeditions, or satisfy a child's wish to sleep under the stars. The backyard camp, like my adoption of its sequence and guidebook format, schematizes the process of making a place, albeit in a familiar locale, to explore the workability and significance of a broader situation. It still gets dark, and we still experience the reverie of the campfire before departing for the known conditions of home.

Susan Sontag wrote that it is impossible to be treatiselike when discussing camp as a sensibility.[1] But taking up camp as method requires a particular kind of treatise that combines guide, manual, and scrapbook.

Such a "fugitive" treatise works between practical didacticism and campfire reverie. The typical camping guidebook departs from the provision of information about particular places usually found in tourist guides and instead outlines a practice and presents possible situations for which the prospective traveler-camper should prepare. In its generic form, the camping guide is not placeless but is "open to all places" and often appends directories and lists of campgrounds specific to a region. The guide also readies the prospective camper to work with the specificity of a place through operations, often a series of objectives and activities. In spite of this open-endedness, the camping guide like the tourist guide does maintain the perspective that the campers will be visitors and strangers in a foreign place. As a result, the guide in most cases remains limited to that which is immediately accessible and visible to the outsider.

Outlining a practice that follows a particular procedural sequence (siting, clearing, making, breaking) and emphasizing necessary activities, the camping guide resembles the manual. This guide-manual coupling complicates the simple application of instructions found in the manual and the information explicated in the guide. The guide aids in seeing and gaining orientation and knowledge—from its Sanskrit origins in *veda*, "to see." The manual, with its "hand" rooted in Latinate *manus*, requires touching and thus gaining skill and knowledge. The format and the critical ground of this combination occur between hand and eye. Like camping, such coordination requires practice and repetition in the context of real situations. The camping guide-manual goes beyond the didacticism of the guide's conveyance of information and the manual's instructional purpose. Present in its comfortable "fit in the hand" and its necessity of being "kept at hand," the portability of the guide-manual object itself allows it to be relocated and its contents to be transformed by a new place and context. The military camping manuals of the Army Corps of Engineers and other units were carried in belt pouches designed to hold documents, food, and other stock. Their attachment to the belt allowed for easy access and protection from the elements. The connection between hand, eye, and camping operations ranges from a tourist's need for immediate roadside orientation and less hurried directions for setting up camp to the soldier's urgent engagement with adverse conditions. In both cases, the guide-manual becomes a heuristic device for place-making. At this intersection of the conceptual and the practical, the procedure of camping informs a distinctive method for occupying and moving within the places around us.[2]

Guides and manuals directly and indirectly related to camping demonstrate diverse modes of engaging particular places. In *Camp Life in Florida*, Charles Hallock combines descriptions of camping techniques with advice on hunting and recreation. Compiled in part from essays published in *Forest and Stream*, Hallock's guide narrates his and other explorers' excursions between 1873 and 1875 in a Florida of which "so little is known." One representative journey within the Florida frontier takes L. A. Beardslee along the St. Johns River from Jacksonville to Enterprise with the service of a "lawn tent" in search of the black bass.[3] Because of the Florida peninsula's relative inaccessibility in the later nineteenth century, the combination of tent and boat was a common way to explore the region.[4] In a subsequent volume titled *Camping and Cruising in Florida*, James Henshall records travels throughout Florida, particularly along the Indian River in his craft called the *Bluebill*. The narrative tells of daily journeys that are structured only by the process of "leaving camp" and returning "back to camp."[5] These accounts, for the most part, assume a vicarious voyage by the reader who will not likely be making camp in Florida's nineteenth-century wilderness. The narratives do, however, document precise connections between camping procedures and specific places.

The growth of North America's road network and the continued development of automobiles resulted in a proliferation of autocamping manuals in the 1920s and 1930s. Elon Jessup's *The Motor Camping Book*, one of the first such manuals, exemplifies the growing connections between a vacationing public and an expanding territory made available for the practice of camping. Documentation of Tin Can Tourist camps by the photographer Ernest Meyer illustrates this variety of campsites allowed for by the expansion and increased quality of the road system, beginning in the late 1910s.[6] Meyer's photographs depict Florida tourists camping in farm lots, alongside roadways, in longleaf pine stands, and in more regulated community campgrounds. Within this context, the purpose of Jessup's book is "to give a practical working knowledge of how to camp out along the way while touring in a motor car."[7] Jessup contends that the motorist who carries a camping outfit and is well prepared for the trip achieves the "greatest degree of travel pleasure and freedom."

"Triumphal Procession" in James A. Henshall, *Camping and Cruising*, 1888. University Libraries, Florida State University, Tallahassee, Fla.

The organization of Jessup's treatise follows the camping process itself but also adopts an elegiac tone that captures the camping spirit of the time. In an early section of the manual, Jessup asks why we motorcamp—supplementing a highly practical objective with a rhetorically philosophical meditation. For Jessup's traveler in the 1920s, it is "the nomadic instinct for a free life in the outdoors [that has run] in our blood . . . for generations."[8] With the affable admonition that "time and space are at your beck and call," Jessup outlines the carefree nature of picking a campsite at the end of each day. One night the camper "may be camped in the yard of a little red schoolhouse, the next in a farmer's orchard, . . . and then perhaps the following sundown finds him setting up his tent in the sophisticated grove of a city park." Jessup also echoes fellow camper-journalists of the time in his emphasis on the importance of "right equipment." Four years earlier in 1917, Emily Post traveled westward in an excursion documented in *By Motor to the Golden Gate*, which became a de facto guidebook for independent women who wanted to explore the nation's road network.[9] Post's accommodations did not include autocamps (she stayed in the finest hotels), but her commentary is not unlike Jessup's reminders, such as "one goes camping to have fun, not to be annoyed." At other times, Jessup describes the functionality of "the cooking fire" but then concludes with a poetic reverie about campfire. He longingly writes, "I must express an extreme partiality for a lingering heap of glowing wood coals."[10]

Many of these newly developed camping manuals follow changes in technology. With the increased availability of trailers pulled by cars in the 1930s, *Touring with Tent and Trailer* supersedes guides devoted to the early autocamper. Kimball and Decker outline what they refer to as "the science of camping" and include sections on planning the trip, pitching tents, making campsites with trailer and tent, and even caring for one's appearance while on the road. One chapter in particular addresses "well-groomed Motor Campers."[11] Technical and social issues are also covered in Wally Byam's *Trailer Travel Here and Abroad*, which served as a general guide for campers and trailer owners and presented narratives of the Wally Byam Caravan Club's trips across the world.[12] Byam, the founder of Airstream Corporation, narrates a kind of proprietary camping science, describing detailed techniques of stowing, hitching, towing, backing, parking, and leveling. Also, using examples of the Caravan Club's camps, Byam discusses how the original camp circle has evolved into the larger-scale "wagon-wheel style" formed by parked Airstreams. More recent guides focus on the less technical, more mythologized, practice of camping as a return to nature and as wholesome entertainment for the family. In *Camping*, Eric Dominy mixes tips on building campfires with techniques for using and unfolding trailer tents—all of which can involve each member of the family. A Public Affairs Committee pamphlet of 1966 proposes camping as an antidote to social problems such as the dissolution of family structure. The goal of "family camping" follows the Public Affairs Committee's mandate of 1935 to "develop new techniques to educate the American public on vital economic and social problems."[13] Presented with a similarly nostalgic quality but without the social directive, Dan and Inez Morris's guide *The Weekend Camper* from 1973 presents the typical procedure for camping: choosing a campsite, making and breaking camp, and building the campfire and the cooking fire.[14]

Preceding these more contemporary examples of camping procedure is Reverend B. W. Gorham's *Camp Meeting Manual* created for Methodist

Airstream rally formation, Auburn, Washington, 1962. Courtesy of Airstream Corporation.

preachers and their adherents.[15] One of Gorham's objectives is to stop the gradual abandonment of camp meetings by adapting them to the contemporary "taste[s] of the people and the spirit of the age."[16] In this revisionist document, the minister also seeks to guard the church by maintaining a degree of doctrinal and thus spiritual control of the meetings, both in terms of formulation and content. By creating the manual, Gorham hopes to prove the utility and, at the same time, to increase the efficacy of the camp meeting event. Along these lines, the portable manual, measuring less than 4 by 6 inches, serves the itinerant Methodist preacher in his "circuit" of stops within the territories defined by the church as "districts." Gorham sees these temporary assemblies administered by the preacher as being as important to the doctrine of Methodism as the permanent churches themselves. A providentially mandated construction, the camp meeting links to the ecclesiastical system of Methodism by establishing a place of spiritual, even mystical, isolation. For Gorham, the camp meeting requires the removal of people from "worldly care," serves as a place where "sublime truths of revelation" are sustained and the "mind of the church may rise," and allows for a break from the "worldliness of summer [in a] singular occasion for conversion."[17] These attributes revolve around the simple act of "going into the woods" and setting up a camp away from the temptations, excesses, and distractions of the "world."

Gorham divides the manual into two parts to demonstrate the camp's connection between soul and body or, in terms of process, the thinking and the making of a camp meeting. A conversation between a skeptic and a believer, ostensibly Gorham himself, defines the first part's structure. The second part, "Practical Observations and Directions," includes technical information about constructing and administering the camp. Introducing the minister's defense of the assemblies and addressing the doctrinal questions outlined above, the first part also serves as a guide for what the participant can expect upon arrival at the camp.[18] In the second part, Gorham outlines the "preparation of the ground" that should occur after the determination of which ministerial circuits will participate in the assembly and where the camp meeting will be sited. This preparation begins by ascertaining the bounds of the assembly and inscribing the circular form. The site is then cleared and graded before the initial siting of the preacher's stand around which the rest of the camp's components will be arranged. Dimensioned to cover approximately twenty-five feet square, the altar is then located in front of the stand; and a "broad aisle" (between seven and nine feet wide) runs on axis with the altar to separate the seating areas of the male and female participants. The circle contains all of the components except for the family and social tents, which are sited tangential to the original geometric layout.

In addition to this rigorous and hierarchical layout, Rules of Order are posted throughout the camp to outline and clarify restrictions on conduct during the meeting. Gorham's notations about the building of the stand, altar, benches, and tents are specific in terms of the procedure and layout to be followed, the materials to be used, and the dimensions to be employed. With its plan dimensions laid out as 12 by 16 feet, the construction of the stand and speaker's platform includes a partition that separates the platform's two-level space front to back. According to Gorham, the rear area is used for lodging and "secret devotions."[19] The specific treatment and dimen-

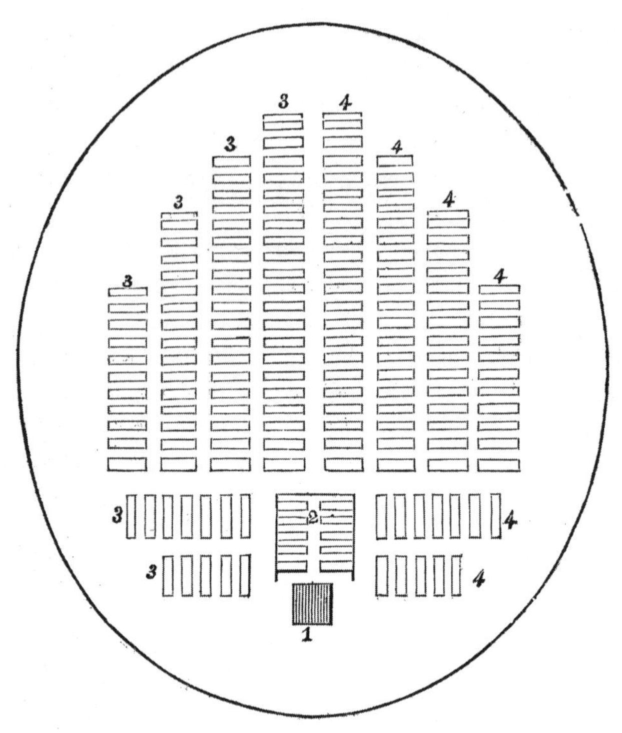

Site plan of Camp Meeting, measuring 14 x 16 rods with (1) speakers' platform, (2) altar, (3) seats on ladies' side, (4) seats on gentlemen's side, and the circle on the outside of which the tents are to be built. B. W. Gorham, *Camp Meeting Manual*, 1854.

sioning of the "book board" on which biblical texts will rest in the upper level of the platform requires that its component parts be properly dressed and planed in contrast to other rougher boards used for the general structure of the stand.

Published during World War II, field manuals for military camps follow a similar sequence of operations in setting up camp but exhibit an emphasis on efficiency and function without the nostalgic or spiritual associations of recreational, social, and religious camping practice. The U.S. War Department's *Staff Officers' Field Manual* (FM 101–10) describes the typical layout of "semi-permanent camps" and more temporary "shelter tent camps." The field manual includes formulas for calculating square footage based on infantry size and on the number of vehicles and animals. The layout of the shelter tent camp occurs along a linear measure of its length and follows a rigid hierarchy of rank and service. Across its main axis, the camp is arranged symmetrically around the commanding officer's shelter, which is at the top and in the middle. Exceptions to the symmetry are the vehicle parking area and picketed animal pens at the bottom of the layout. The guide also dictates the "degree of dispersion" of tents and vehicles within the bivouac area of the temporary encampment.[20]

Also prepared by the U.S. War Department, the *Engineer Field Manual* includes more extensive operational procedures for siting and constructing camps. This field manual presents four categories of troop shelters—based on the duration of the camp and its degree of permanency. For each category, the manual addresses the occupation of places before, during, and after battle along with the added variables and the varying degrees of permanence and privacy. The types of camp shelter occur in the order of decreasing temporality—a hierarchical framework of encampment. Bivouacs are sites "in which troops rest on the ground covered by shelter tents or hastily improvised shelter." Camps occur where "cover is provided by tentage more elaborate than shelter tents." Cantonments serve as semipermanent camps "in which shelter is provided by buildings erected for that purpose." And billets occur where "shelter is provided in [preexisting] public or private buildings."[21] Highlighting the manual's combination of pragmatic and strategic objectives, paragraph 158 outlines the "selection of camp and bivouac sites" based on comfort and convenience as well as the tactical advantages afforded by a particular location. Another U.S. War Department field manual, FM 20–15, specifically addresses erecting tents within military camps. This manual includes guidance in siting, pitching, trenching, striking ("breaking"), and folding tents. A list of rules governs the selection of site—"do not camp at the base of a cliff" and "choose level ground." In contrast to the larger scale issues of siting a camp, the section on pitching tents presents a sequence of directed procedures from "divide the tent pole sections into four parts" to "stake out the side guy ropes on tent pins."[22] The field manual combines tactics and comfort and efficacy and intuition in its procedural directions scaled from regional-environmental forces to the soldier's bodily movements required to assemble a tent.

The guide-manual format narrates and frames transitory events. The architect John Hejduk's sketchbooks and projects record transient places of experience and imagination. Hejduk's urban masques, carnivals, and expositions move between particular cities and places. Documenting journeys associated with Venice, Berlin, and Russia, these works can also be read as travel diaries and retrospective guidebooks. Hejduk writes of the fleeting and semipermanent nature of the constructions and events in these manuals for urban exposition: "I have established a repertoire of objects/subjects and the troupe accompanies me from city to city, from place to place, to cities I have been to and to cities I have never visited." The fictional masques staged in Hejduk's work are impermanent in their physical nature but are enduring as highly personal guides to real and imagined cities. Hejduk has developed a method of place-making: "I believe that this method/practice is a new way of approaching the architecture of a city and of giving proper respect to a city's inhabitants. It confronts head-on a pathology."[23] The accompanying troupe deploys itself, quite literally camping, in each successive city to interrogate the particular urban conditions—from public squares in Vladivostok to varied urban spaces in Berlin. Such a manual's procedural and speculative construction reflects, records, and negotiates the residual nature of fairs and camps.

Just as Hejduk's speculative oeuvre accumulates the meanings of places visited by the architect, the *Guide to the Southernmost State* pulls together disparate fragments to form a lyrical and didactic compendium. The guide produced by the Federal Writers' Project for Florida includes information on the state's camps and campgrounds in the 1930s and represents a mode of recording information that connects with the process of camping itself.[24] Although primarily structured by its twenty-two tours, the guide can be read as a collection of fragments each of which is a separate story within the region's larger narrative.[25] Like camping, the Florida guide does not require a linear reading—its itineraries crisscross and can be cross-referenced with equal weight given to fact and myth. As an alternative history of the places, the text guides visitors and tourists from outside the state and serves as a reference for those already familiar with the Floridian environment. The *Guide to the Southernmost State* is a collection of epigraphs, block quotations, transcribed speech, and the text of sign-postings in which these fragments are sewn and woven together. It is this patchwork composition that relates to the bricolaged vehicles, dwellings, and grounds of the Florida camps.

As an assemblage, the camping space parallels other works that define and document place-making practices. More consciously wrought as an assembly of fragments, *Let Us Now Praise Famous Men* is structured as two books that operate as guide and manual. If the first book serves as the contextual siting of the story and actors, then the second book can be understood as the assembling and disassembling of the story's action.[26] To use the camping metaphor, this second part entails the story's clearing, making, and subsequent renewal of breaking. The narrative structure moves from made things like shelter and clothing to an analysis, literally a "breaking," of the place.[27] Within this loosely structured format, Agee summarizes the desire that his writing should sufficiently present the fragmentary situation of the project's subjects: "If I could do it, I'd do no writing at all here. It would be photographs; the rest would be fragments of cloth, bits of cotton, lumps of earth, records of speech, pieces of wood and iron, phials of odors."[28]

Like John Hejduk's sketchbook-manuals, Ruskin's museum-map construction circumscribes a version of the guide-manual that allows for both explanation and exploration. In *St. Marks Rest*, John Ruskin uses an open-ended method to assemble a fragmentary, yet "truer than you have heard hitherto," account of the city of Venice. Mirroring Ruskin's own peripatetic method inclined to digression, the guide combines historical, architectural, mythological, and legendary accounts to present the "book of a nation's art."[29] Ruskin does not attempt to traverse the entirety of the city but instead finds the history of Venice written into its details; and his understanding of the city is not linear but topological, as he folds time and space in the interest of his narrative journey driven by combining the specific realities of place and the imagined "dream-space" of his own personal historical discourse. Within the "contracted world" of museums and catalogues, Ruskin synthesizes the museum and the map.[30] In this sense, the guide is ordered by Ruskin's personally curated collection in the given reality of the Venetian landscape.[31]

Manuals produced for the World's Columbian Exposition further characterize the structure and experience of events within this "contracted world" of place as camp. Manuals for experiencing the World's Columbian Exposition of 1893 took the form of guides and catalogues. Evoking the assumed impermanence of camps, Richard J. Murphy's *Authentic Visitors' Guide* notes the temporary nature of the event: "All buildings, with probably one exception, to be decided on after the close of the Exposition, will be removed from the grounds within six months after the gates are shut in October."[32] The guide memorializes the fleeting experience of the fairgoer and contends that the text itself will serve as a vehicle for continued edification and imagination. Within the condensed format of its seventy pages, the *Authentic Visitors' Guide* sought to "furnish, in brief and attractive form, all information required by the stranger relating to the Exposition and the city of Chicago."[33] The guide's system of classification functions as both index and catalogue, containing "1,000 classified subjects" and taking the form of a "Finding List."[34] Designated as a "concise method of locating exhibits in all buildings," this list of immense diversity includes articles from Academies to Zinc, from Aromatic Substances to Axle Grease, from Crystallography to Immigration.[35] This classification of sites and subjects parallels the camping guide's presentation of places to see. The *Authentic Visitors' Guide* helps the visitor navigate the artificial geography of the Exposition's terrain, just as the camping guide seeks to orient the traveler and anticipate the camper's context.

Other Exposition guides sought to locate the visitor not by subject and coordinates but through the historical and cultural milieu of this event space. Lacking overt systematization in its standard index, the *Official Guide to the World's Columbian Exposition* serves as a handbook with extensive background information about American progress, the city of Chicago, and the fair itself. In this case, the fair becomes a component within the larger political territories of Chicago itself and within the historical and cultural territory of increasingly industrialized American production and its perceived societal progress.[36] The *Official Guide*'s didacticism is apparent in the compiler's introductory encouragement of the visitor "to study the accompanying map. This is an absolute necessity to one who would not travel aimlessly over the grounds and who has a purpose beyond that of a mere curiosity hunter."[37] Echoing the fair's goal to endow visitors with a "liberal education," the guide assumes that visitors seek the "enlightenment" that the fair and the guide can provide as a result of world progress in the arts, sciences, and industries.[38] One other instance of the Exposition's documentation is the thousand-page *Official Catalogue of Exhibits*, an exhaustive compendium of all the exhibits within each department.[39] Each section includes a brief description titled "Key to Installation," explaining the internal arrangement of the exhibitions and providing precisely detailed building plans.[40] Illustrating each exhibition's layout with layers of text, numbers, and partitions, the plans serve as both map and catalogue in their combination of scaled floor plan and diagram.[41]

These floor plans are campsites through which the visitor travels, lingering to learn of a distant place or to imagine a possible future. Each of the Exposition's temporary buildings becomes a massive tent,[42] housing a protected ground in which countries might "camp out" during the Exposition. It is not surprising, then, that the Transportation Building displayed the Gypsy caravan, predicting the rise of an autocamping public after the turn of the century.[43] Immediately outside of the Exposition's main area, other camps supplemented these temporary grounds of display. Buffalo Bill camped in his Wild West exhibit, and National Guard troops set up camp on the grounds of Chicago's federal buildings. An extension of the Exposition's main grounds, the Midway Plaisance included its own set of camping installations.

Completing the guide-manual interaction, the scrapbook can also be linked to camping practices and places. The methodology of assembling a scrapbook resonates with some of the procedures found in Ruskin's composition of the museum-map. The scrapbook is a tourist "memory book" that captures recollected experience but can also serve as a reference for future tours. Such an intersection of memory and geography occurs as a personal recording of events pasted within a journal's blank pages.[44] While a diary may remain undisclosed to a public reading, the scrapbook records and preserves fragments (pictures, newspaper clippings, tickets, and other scraps) for both private and public reference. Like the travel slide show, the scrapbook is prepared and constructed to anticipate a potential performance or "reading." Within the context of the guide-manual coupling and its implications for a camping practice, the scrapbook completes this idea of "camp as method"—a rethinking of places experienced.

In the Florida State Archives, we find the confluence of camping and scrapbooking in a series of scrapbooks that narrate the story of the Tin Can Tourists of the World, arguably North America's first autocamping society. Composed by two group members, Ray and Mary Levett, the scrapbooks document the functions of the group, the various campsites where meetings were held, the evolution of the trailers used, the automobiles that towed them, and the activities and amusements enjoyed by the members.[45] As historical documents, these scrapbooks serve as manuals for retrospectively interpreting the camping practices of the Tin Can Tourists. Moreover, the scrapbook as a methodology follows what Ruskin called a "caravannish manner" in which seemingly disparate items are pasted together, linked, and juxtaposed.[46] This documentary method was used in an eminent British Broadcasting Company series called *Scrap-book* in which a new history of place was narrated through oral fragments.[47] As an architectural research method, this combination of manual, guide, and scrapbook is proposed as an open-ended "process manual" related to camping practice.

Chapter 3

CLEARING CAMP

Camp's Field of Meaning

> I admit it's terribly hard to define. You have to meditate on it [camp] and feel it intuitively, like Laotse's Tao.
>
> —CHRISTOPHER ISHERWOOD (1954)

If siting is the forum for outlining a camping practice, then clearing defines camp itself. In the former, the guide-manual-scrapbook frames the process remotely. Clearing camp then enlists the multivalent possibilities of site that make room for camp's place-specific meaning. Given camp's inherent flexibility, its definition not only is denotative but also must include its wider connotative field. The constellation of meaning found across camp's connotations embraces the paradoxical coincidences of permanence with temporariness and dispersal with collection. Camp is first of all a field, usually a level field, sometimes a battlefield or the grounds for a tournament. In contemporary Spanish, *campo* refers abstractly to the country as opposed to the urban and more tangibly to the countryside, which is the open field of the country. The Spanish word *camping* has come to designate the campsite, and *campamento* is a less organized collection of tents. *Campo*'s Latinate origins are found in *campus,* referring more precisely to the level field.[1]

In ancient Rome, the Campus Martius was a multipurpose leveled field. As is eponymously suggested, the Campus Martius was the "field of Mars" dedicated to the god of war. During Rome's republic period, the Campus Martius lay outside the city on its northwestern limit. The multiuse field functioned primarily as a place for military drills in the spirit of its namesake, but the Campus also became a place for games, athletic practice, simulated combat, and public assemblies of citizens and religious adherents. The field's adjacency to the Tiber River and its low-lying elevation made it subject to frequent flooding and necessitated that the activities and events be temporary or short-lived. In 54 BC, the Roman government initiated flood-control projects to counter the effects of the periodic flooding. Emperors Pompey and Caesar added a theater, colonnade, assembly hall, and a new forum with a temple between the Campus and the city proper. Augustus and Agrippa continued the urbanization with greater expansion of building projects and facilities. Along with an array of games and altars dedicated to minor and sometimes marginalized gods, the Athenaeum, Hadrian's school of liberal arts, was reputedly built in the northern Campus Martius. With the perfection of flood control, the Campus became the urban center of medieval Rome and was later documented by Giovanni Battista Piranesi in a series of mid-eighteenth-century engravings known as *Il Campo Marzio dell'Antica Roma*. In Piranesi's *campo*, the buildings and monuments from Giambattista Nolli's map of Rome are layered in a fanciful assemblage of urban space and historical events.

As its Latin designation and the historical coincidence of the Campus Martius and Hadrian's Athenaeum suggest, camp also relates to the "open field" of the college or university campus. Campus planning at the nation's frontiers, beyond urban confines and distractions, has paralleled the American tradition of camping and thinking, from the Emersonian Philosophers' Camp to Chautauqua's institutional network through more institutionalized university systems. In some cases, camps precede a campus. In one case, the campus of Ruskin College in Florida was founded in a turpentine camp. In other situations, campus subsumes a neighboring military camp, as with Rutgers University's absorption of the adjacent grounds of Camp Kilmer. And the design of Thomas Jefferson's University of Virginia campus resonates with the typological form of the idealized military camp's layout.[2] In the southwestern United States, Frank Lloyd Wright's Ocotillo Camp served as the background for his founding of Taliesin West. Wright and his students made their camp in the late 1920s and 1930s, while the site for Taliesin was chosen and construction was carried out.[3] The adaptation of previous camps' form and infrastructure reaches a height of pragmatism during the post–World War II expansion of university campuses. To accommodate the

Stereo photographer Charles Seaver, Jr., in his steamboat-barge along the Ocklawaha River in Putnam County, Florida. The vessel served as a mobile campsite with tent and darkroom for developing his photographs. State Library and Archives of Florida.

influx of students, Rutgers, like many of the nation's universities, reused the decommissioned military buildings from Camp Kilmer. This rapid fusion of the shelters, grounds, and organizational systems of camp and campus stands as a key moment in North American campus planning, the scalar and ideological shifts of which are still being felt today.[4]

Just as the development of the American campus derives in part from the parallel construction of camps, a related meaning from camp's Latin origins connotes the less historical and more conceptual sense of "watery camp." This use of camp refers to the surface of the sea and derives from *caeruleus campus* and *campus latus aquarium*.[5] As an imagined site for the camping experience, the surface of the sea accommodates a web of forces and paratactic conditions in which characteristics and qualities are flattened within the horizontal expanse.[6] In her account of a caravan trip organized by Airstream's Wally Byam, Lillie B. Douglass echoes this camper-mariner confluence in her view of the African landscape's watery camp: "the world seemed a watery blur, without end, and we afloat in it."[7] In this case, the desert is transformed into a series of mirages, or seemingly real sheets of water across the landscape.

These improbable oases lie at the confluence of the real and the imaginary within the particularly disorienting setting of the oceanlike desert. The Bedouin camp, then, serves as a transitory location within the real-unreal environment of the Arabian desert. Echoing Douglass's experience, Gertrude Bell traveled through the Near East in 1892 and later, as Great Britain's Oriental secretary to the high commissioner in Baghdad, helped form the modern state of Iraq. In her narrative and documentary writings, she noted the seemingly impossible coincidences of camp and mirage: "We saw tents with men beside them pitched on the edge of mirage lakes and when at last we actually did come to a stretch of shallow water, it was a long time before I could believe that it was not imaginary."[8] The fluidity of this setting in which water, air, and sand merge proved perplexing to early surveyors of the region and connoted a sublime inflection of the infinite and the immeasurable.[9] In the desert, traditional methods of surveying such as locating geographic-topographic features and positions have to be transformed into more flexible procedures to understand less immediately visible data through such means as astronomical observation and learning to read the desert's void along with its more subtle topographic qualities. Distances collapse, and scale oscillates between the miniature mountain and the immensity of sand granules.

Camps are thus associated with the veritable oasis formed by subtle variations of the ground. In a segment typical of her descriptive style, Bell writes: "we are camped to-night in what is called a valley. It takes a practiced eye to distinguish the valley from the mountain, the one is so shallow, the other so low. The valleys are often two miles wide."[10] The features of the desert are the all-encompassing aspects of the unreal made visible as mirages and the infrequent and subtle anomalies of sand and stone within the sublime emptiness of the desert:

> Though we were riding through plains which were . . . to the casual observer almost featureless, we seldom traveled more than a mile without reaching a spot that had a name. In listening to Arab talk you are struck by this abundant nomenclature. If you ask where a certain sheikh has pitched his tents you will at once be given an exact answer. The map is blank, and when you reach the encampment the landscape is blank also. A rise in the ground, a big stone, a vestige of ruin, not to speak of every possible hollow in which there may be water either in winter or in summer, these are marks sufficiently distinguishing to the nomad eye.[11]

Similar to the *mattang*-mariners of the Marshall Islands discussed previously, the Bedouin's navigation of these attributes through careful observation and translation of the real and the unreal allows for the transient positioning of camps within the ephemeral landscape.[12]

Camp's etymological derivation is not typically taken beyond its *campus* Latin origins.[13] However, in Greek texts, descriptions of camps and campsites include Greek variants of the term *chora*. This lexical connection between camp and *chora* is most evident in Homeric works. For example, in the *Iliad*, Hector makes camp in a space cleared of the dead: "Then did glorious Hector make an assembly of the Trojans, leading them away from the ships beside the eddying river, in an open space (χωρος) where the ground showed clear of the dead."[14] In this use translated here as "open space," χωρος (or *choros*) is also a "piece of ground," or "a place."[15] The specificity of siting on a particular section of the ground is qualified by the related term χαμαδις which denotes directionality and positioning—both *toward* the ground and *on* the ground. In Homeric texts, such a place is notable as a site for speech-making. From the "open space," Hector delivers a rousing speech to the assembled Trojans. The camp as *choros* later serves as the site for a series of speeches from Hector's men: "so they went through and out from the trench they had dug and sat down in an open space where the ground showed clear of dead men fallen."[16] In each case, after speeches are made, campfires are lit and food is prepared. In these references, siting camp includes both a clearing of the dead and a ritual of making speeches at the outset of occupying and assembling within the open space. The cleared open space of the Trojans' temporary military camp, as *choro*, makes a linguistic connection with Greek *chora*, which is both space and place.[17]

Camp's military connotations include a temporary lodging for troops but also a semipermanent station for accepting those troops. Furthermore, camp refers to a body of troops moving together. The encampment, inherent within the military definition, sets up a "city within a city": "camp is also used among the Siamese and East Indians, for a quarter of a town assigned to foreigners, wherein to carry on their commerce. In these camps, each nation forms itself a kind of city apart, in which their store houses and shops are, and the factors and their families reside."[18] Such political and economic enclaves can be compared to the modern embassy. In Australia, the Aboriginal Tent Embassy served as an ad hoc office to represent disenfranchised aboriginal inhabitants and as a form of political protest of Land Rights. Started on January 26, 1972, the assemblage of tents that made up the embassy occupied the grounds of the Provisional Parliament House in the central urban area of Canberra.[19] The building bylaws at the time of the camp's inception did not expressly disallow temporary occupations of Canberra's urban spaces. Six months passed before legislation could be passed to prohibit such encampments and certify the original tent embassy's demolition.[20] Following this destruction, the Aboriginal Tent Embassy was reestablished at regular intervals until 1975 and intermittently until 1992. Having established a greater degree of permanence since 1992, this encampment was registered by the Australian Heritage Commission on the National Estate as the first Aboriginal Heritage Site.[21]

Variants of camp connote a body of adherents to a commonly held doctrine or theory or a defensible position of beliefs, protected by a veritable "army" of arguments or facts. As a position from which ideas are defended, camp takes on the quality of an idea itself and thus describes mental activity and imagery. Camp can also be understood as a field of investigation, discussion, or debate. Such derivation from the physical reference to the field of play or combat yields the more conceptual and abstract notion that a camp is an epistemological area "opened" for debate. The architect R. E. Somol takes up the idea of camp as a confluence of old and new—the retrospective view and the production of the manifesto. Somol argues that the contemporary practices of such architects as Rem Koolhaas rely on the postmodern subversion of the logical and the sometimes arbitrarily lyrical. The resulting "camp of the new" defines a measured avant-gardism that experiments with urban forms while critiquing modernist ideas of inevitable, inviolate progress.[22] In terms of architectural method both old and new, camp describes an idea and inscribes the schematic zone for its development—its "becoming." Consequently, camp relates to method both in its concrete procedures and through its more abstract mental activities of making place.

Although distinct from "camp as idea" but within this conceptual understanding, camp also includes a sensibility defined through intensive intellectual perception and mental responsiveness to situations. In her 1964 essay "Notes on Camp," Sontag outlines camp as a third sensibility that offers supplementary standards to those of high culture and the avant-garde. By titling her essay "Notes on," she indicates that the discussion of camp requires an alternative methodology, framed to allow for both pathos and objective distance. This notational mode, which she refers to as "jottings" in contrast to the linear logic of the formal essay, places her own argument as a commentary within camp. Sontag discusses this necessary paradox: "To name a sensibility, to draw its contours and to recount its history, requires a deep sympathy modified by revulsion."[23] Freed from the standard dichotomy of literal and symbolic meaning, the production of camp operates in the differential space between the "thing as meaning something, anything, and the thing as pure artifice."[24] Sontag's treatment of camp simultaneously identifies this sensibility and transforms the oppositional structure of an objective and subjective framework to create her own critical space—to "set up camp."

Camp joins this idea of a third sensibility with the paradox of a mobile fixity and an unstable permanence. This combination consequently elicits a nonnostalgic rediscovery of ways of living that have been marginalized or forgotten by a dominant culture. As understood by Sontag, the "fugitive sensibility" of camp is a set of values supplemental to societal standards. This paradox of a shifting system of values disallows the possibility of completeness as promoted by high culture's oeuvre in favor of a lingering cultural fragmentation. The difficulty of siting camp lies in the complications of the "utopic impulse" and its unique set of values made "fugitive" by the inherent mobility of camp. This mobility is reflected in the multiple possibilities of siting camp semantically as well as situationally. Sontag notes camp's flexibility of meaning in its usage as a verb to describe a "mode of seduction." Not resulting in complete instability, seduction does imply that one is always being "led away." Such attractions to a person, place, or climate necessitate habitual movement. Situationally, the siting of camp operates as pure process—in fact, its stability is in this consistency of movement, a cycle of seduction and of being led away to the next place and the next campsite. Sontag further clarifies this circumscribed definition of camp: "What it [camp] does is to offer for art (and life) a different—a supplementary—set of standards."[25]

As an architecture sometimes categorized within vernacular production, camps denote conscious, place-specific decisions about making architecture. Sontag's notion of camp resonates with everyday architecture and its counterpart vernacular architecture. Neither architecture nor the everyday is characterized by naiveté. Like a methodologically driven camp, "the making of architecture is a highly conscious, indeed a self-conscious act." And the everyday should not be confused with a "sugary and debased notion of the vernacular" and the concomitant nostalgia for "some state of original purity."[26] Like Sontag's characterization of camp as supplemental and definition-defying, the everyday and its related camping phenomena resist co-option.

As a supplementary practice, camp also connotes the performative. This meaning of camp emerges as early as 1909 in J. Redding Ware's *Passing English of the Victorian Era* in which camp describes "actions and gestures of exaggerated emphasis."[27] Such embellished behavior was associated with homosexual mannerisms and taste starting in the 1930s and 1940s. Mark Booth traces this derivation of camp to the French term *se camper*, which means to "act broadly and histrionically, to be expansive but flimsy."[28] Method-acting, theatricality, and exaggeration can also be included within the broader meaning of *se camper*. Though gay culture may have been at what Sontag calls the vanguard of defining camp taste, such aesthetic and even philosophical sensibility cuts across time and culture to include society as whole.

Camp as an idea has broad implications for aesthetics and philosophy. In the novel *The World in the Evening*, Christopher Isherwood digresses from his autobiographical account to write about the difficulty of defining this aspect of camp, requiring meditation to "feel it intuitively."[29] In Isherwood's formulation, camp combines a general applicability in philosophical demonstrations with the potential for a highly subjective and individual method. This idea of camp approaches Lao Tzu's "the way," particularly in the Chinese philosopher's ideas about water. Recalling the watery camp of desert and sea, Lao Tzu writes, "because water excels in benefiting the myriad creatures without contending with them and settles where none would like to be, it comes close to the way. In a home it is the site that matters."[30] The fluidity of water and camp allows for an approach to the "way" that remains labile and infinitely adaptable. This inherent flexibility and lightness of camp makes for a habitability of the seemingly uninhabitable and a suitability of the apparently inappropriate. This dialogue between possible and impossible sites, as well as nonsites and sites, reflects camp's paradoxical home-away-from-home and homely *unheimlich*. Given this complexity and ironic turning, camp remains "terribly hard to define."

If camp as the "way" requires a meditation on such paradoxes, then camp as an aphoristic style matches Nietzsche's practice of rumination. The aphorism waits for meaning from the application of external forces.[31] As places of waiting and relaying both intellectually and materially, camps work aphoristically within a Deleuzian smooth space: "the form of exteriority situates thought in a smooth space that it must occupy without counting, and for which there is no possible method . . . but only relays, intermezzos, resurgences."[32] Camps are the relay stations for nomadic thought, and the meaning of places comes in part from these exteriorized practices of thinking. Camping as thinking.[33]

Camp can also be defined as a "weak" architecture. In its tangential quality, weak architecture allows for an open epistemological system and a productive, though unstable, ground of reference. Opposing the weak and the peripheral to the dominating and the centralized, this architecture arises not from foundational positions or modernist ideals of progress and stability but from recollections, events, resonances, and fragments: "contemporary architecture, in conjunction with the other arts, is confronted with the need to build on air, to build in the void."[34] The "weakness" reflects conditions of contemporary culture in which the reality of modernist thought must be addressed to avoid a crisis of coherence if the rigors and technologies of modernism are completely rejected. The event, along with its attendant intensities, becomes a dialogic reconditioning of modernist rigor within the pluralism and highly circumstantial nature of place and time.[35] Building on air and within the void requires a new way of clearing and thinking about ground, perhaps visualized as a diagonal, oblique relation between components.[36] And closer to the idea of camping, weak architecture is a grounding without ground, suggesting that every interval requires a newly formed goal and its grounding.[37]

In critical analyses of art, culture, and society, the camp serves as an archetypal site for shelter. Gottfried Semper relates clothing and camping in the historical lineage of housing the body: "the art of dressing the body's nakedness . . . is probably a later invention than the use of coverings for encampments and spatial enclosures."[38] If architecture precedes clothing, then the camp gives rise to Semper's four elements of architecture and the principles of "dressing." The four generative elements are hearth, roof, enclosure, and mound, with their concomitant practices of ceramics, carpentry, textiles, and masonry. The hearth clearly links to the campfire, and the roof, or primary enclosure, can be associated with the tent. Emphasizing the role of the tent, readings of Semper often ignore the importance of the camp as the nexus in which the "elemental operations" are acted out.[39] We can understand the development of the camp as an active response to necessities of shelter and communal life as well as a unique process and method that eschews both permanence and codification.

The responses of these camping procedures follow two thematic threads that Semper outlines in his explication of the four elements. Before beginning an ethnological account of the four phases in the development of building practice, on his way to proving the existence of a polychromic architecture, Semper outlines two types of camps in which all themes and practices originate—the open and the closed. The former type of camp, in which the hut form predominates and the roof is emphasized, develops as "free and asymmetrical groupings" characterized by an irregularity, resulting from isolation and a lack of property rights. In agrarian cases, this type of camp is a product of the communal efforts to conquer nature. The openness of this camp is further qualified by its slow growth under a "native sovereign" and its expansion by "development and improvement of the simple and the small."[40] With societal competition and conquest comes the prevalence of Semper's second type—the fortified camp. Sharing the generative and thematic aspects of the four elements, the construction of the camp does, however, require a defensibility and closure antithetical to the hut-based asymmetrical settlements of the agrarian-nomadic societies. The fortified camp yields the typology of the court building with its "regularity, clarity, convenient planning and strength."[41] In contrast to the native sovereign, vassalage necessitates rapid construction of the camp. Subsequent building imitates the vassal's infrastructure in which the camp of the leader serves as the model for smaller-scale developments down to the smallest units of the lowest class. By explicating this evolution of the camp as a nomadic, agrarian, and feudal construct, Semper defines the four elemental generative forces for modern revisions to the understanding of shelter and place.[42]

In other formulations, the campground is a totalizing and utopic scheme. The campground's foundational and communal characteristics define typologies of encampment.[43] As a place for the synthesis of these main topics, the camp "was a site of two intertwined developments: the birth of architecture and matriarchy."[44] Greek architecture evolved along an initially domestic lineage from camp and campfire to *megaron* and hearth and finally to temples of Hera in the seventh century BC. Betsky finds the form of the campground as the "first architecture of human culture," inscribed within the mandala. Camps also reflect the form and form-making sequence of the mandala. The ritualized camp configuration becomes a "porous circle marked out in the landscape that turns the continuum of nature into a place where artifacts are created."[45] Betsky does not go further into this connection, but as wheels of time, the mandalas map temporality and spatiality in the flattened geometry of circle and square and reflect other aspects of the camping process. Their construction in sand by itinerant Tibetan Buddhist monks reflects the making of camp. The creation of the sand mandala follows a cyclic process similar to the archetypal camping procedure. Buddhist monks typically begin by consecrating the site with chants and mantra recitation. The monks continue with the drawing of the lines and constructing the mandala with grains of sand that pass through the chakpur funnels. The ritual concludes with a consecration ceremony in which monks dismantle the mandala and ultimately disperse the sand in a flowing body of water. Like the temporality associated with camps, the Tibetan mandalas endure for a limited time. The mandala diagrams the cosmos and serves as a meditative guide both in making and observing the heavens.

Rather than a static form of settlement, camp can be understood as a dynamic cultural construction. Betsky's definition of camp follows this simultaneously open-ended and precise frame: "It is horizontal, not vertical. It is neither womb nor projection, but a place of gathering and definition. It is not hierarchical, and it is difficult to define as a separate place: It does not have an inside and outside so much as it has a center and a periphery."[46] Like the tent, camp is the physical and metaphysical weaving that makes sense of the world. In more contemporary domestic spaces, the woven camp interrogates modern relations of inside and outside with inversions of private and public. This "indoor-outdoor" aesthetic influences the relation of the home to media. Lynn Spigel argues that the "mobile privatization" of the stationary television set linking the home to the world becomes the "privatized mobility" of the portable electronic device along with interior-exterior spatial connectivity of the "modern" home. Such domesticity allows for the secure experience of the rustic and the outdoor. A 1966 *Ladies Home Journal* article titled "Indoor Camp-out" identified the living room floor as the "ideal campsite" and proposed that a new line of sleeping bags made for an indoor-outdoor "adventure for the whole family."[47] These backyard and living-room campouts occur in the domesticated wilderness of the suburban lawn and garden or within the protected zones of the family's house, where the televisual flicker supplants the flame of the campfire.

Forming Camp: Clearing a Place for the City

A camping area is a form, however primitive, of a city.
—CONSTANT NIEUWENHUYS (1966)

Clearing for camp works from the domestic scale of the living room to the urban scale of the global city. As the avant-garde architect Constant notes, camping quite literally forms a city—a built environment with all of its material and organizational necessities as well as the programmatic needs of its occupants. The evolution of the camping form can be tied to the planning and growth of our urban environments generally and our places of dwelling specifically. This clearing for human occupation occurs at the scale of the city and the body, in the large territories of development and within the more intimate mobile dwellings that make up camping communities. Camping form, often arrived at empirically and without professional or official mandate, is a populist precedent to many of the urban and suburban types prevalent in the contemporary landscape. Just as some camps presuppose the academic campus, many camps have become cities. In some cases, this formal development comes about from a freezing of temporary occupation in a more permanent structure as with Michel Butor's reading of Istanbul's incomplete solidification of awnings turned into roofs. In other cases, vernacular systems have generated forms that are then interpreted for their potential use in more modern and institutionalized applications. One trajectory of camp's generative form moves from spiritualist-secular to social-demographic to vernacular-avant-garde.

Clearing for Play: From Camp Meeting to Seaside Urbanism

> A camp meeting however is a thing so outrageous in its form and in its practices, that I resolved to go to one.
>
> —BENJAMIN LATROBE (1809)

With a history that extends back to 1799 in central Kentucky, the camp meeting was originally comprised of an informal grouping of tents sited in rural areas and woodlands. B. W. Gorham identifies the origin of the Methodist camp meeting with the tour of William and John McGee. The brothers "tarried to attend a Sacramental service . . . about to be held" along the Red River in Kentucky. Because of its unexpected popularity with the region's inhabitants, the service assembly was continued and became known as the first camp meeting, from which other meetings in adjacent districts were spawned.[48] Itinerant Methodist preachers arranged, organized, and performed the siting, construction, and sermons of the camp meetings that were designed to add converts to the Methodist faith and to renew those who had been spiritually lost. To maintain this loosely held congregation, the itinerant preachers followed a circuit of locations that created an "other world," or "other place," transcending the locative worldliness of permanent homes.[49]

In its later permutations, the Methodist camp meeting became a semipermanent place for religious experience. In the case of Wesleyan Grove, the city of tents evolved into an agglomeration of cottages by 1859. With this increased permanence, the entrepreneurial development of Oak Bluffs on property adjacent to Wesleyan Grove's Cottage City in 1866 challenged and

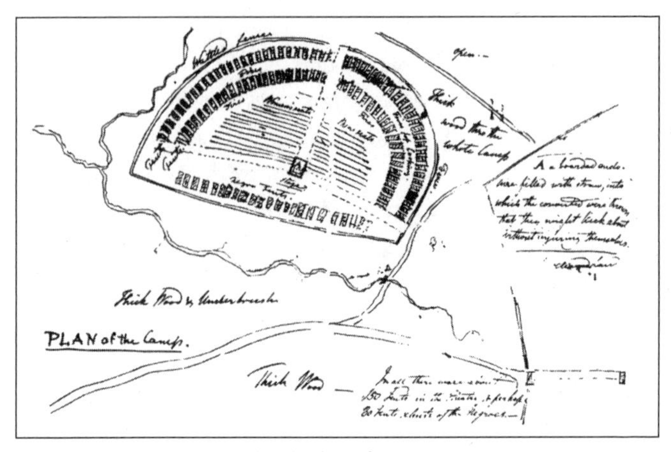

Benjamin Henry Latrobe, sketch plan of camp meeting in Virginia, 1809. Courtesy of the Maryland Historical Society.

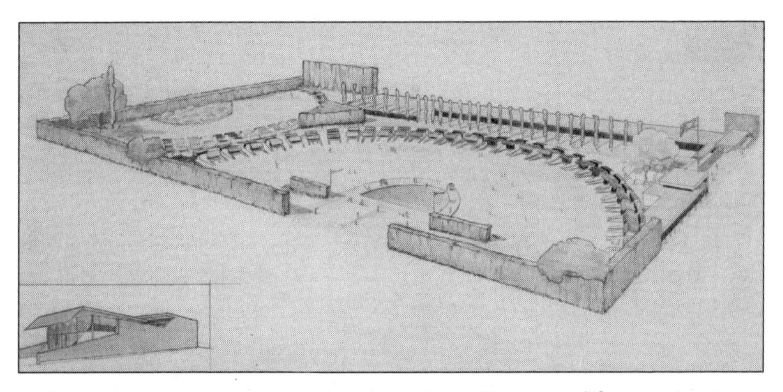

R. M. Schindler, A. E. Rose Beach Colony, Santa Monica, California, 1937. R. M. Schindler Collection, Architecture and Design Collection, University Art Museum, University of California, Santa Barbara.

partially reversed the original focus on religious transformation. Although reflecting much of Wesleyan Grove's scale, spacing, and road layout, Oak Bluffs became a middle-class family resort in which the spiritual mission was supplemental to the community's social life. This transformation from spiritual to primarily secular foreshadows future adaptations of the camp meeting's form and ideology of community.

The mediating potential of the campground continued to develop with

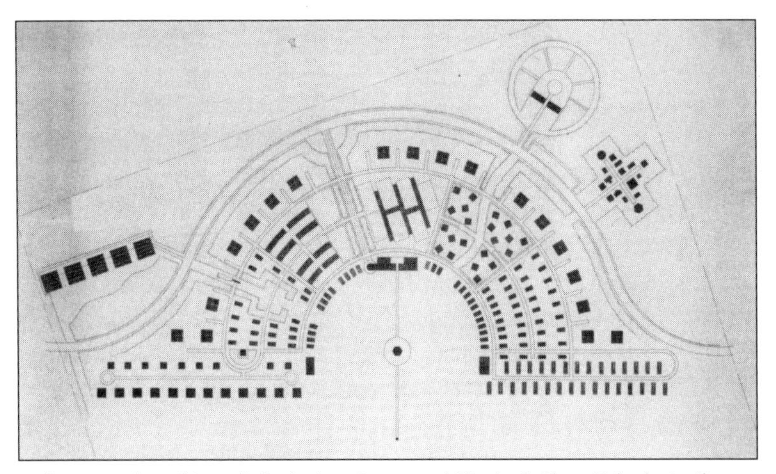

Preliminary plan of Seaside by Andres Duany and Elizabeth Plater-Zyberk, April 1979. Courtesy of Duany Plater-Zyberk and Company.

the spiritualist camps later in the nineteenth century. In the specific context of the Spiritualist belief that the physical world of the living can communicate with the spiritual world of the dead, the campground space itself mediates between the secular and the spiritual, the artificial and the natural, and the permanent and the temporary. We might carry this further and say that the camp takes the form of the Spiritualist "medium," the person acting as the conduit for supernatural forces and disembodied spirits. In this sense, camp is physical and ephemeral ground as well as a less tangible intellectual environment and medium of physical-metaphysical communication. Built form and spiritualist ideology connect through nonhierarchical organization, ordered more by ritual and practice than by the form itself.

The Spiritualist camp at Cassadaga in Lake Helen, Florida, reflects this idea of the campground as a mediating zone between the physical and spiritual worlds. Formed in 1895, the fifty-seven-acre Spiritualist camp grew to include fifty permanent houses within the campgrounds and approximately twenty outside the gates of the camp. George Colby, the founder of Cassadaga, was directed by his American Indian spiritual guide, Seneca, to follow "'a footpath . . . through the deep forest'" and subsequently made a journey from Eau Claire, Wisconsin, to central Florida, where the rolling terrain reminded him of his "boyhood home in western New York," inspiring him to found the Spiritualist camp.[50] Though the scale and particular practices of each camp vary, Cassadaga's mission and philosophy are related to camp assembly grounds established in the northern United States for combinations of education, spirituality, and leisure. These northern precursors to Cassadaga include the Chautauqua Institution, organized in 1874 in Chautauqua County, New York, and the Lily Dale Assembly, established by the National Spiritualist Association in 1879 and incorporated in 1893. Mirroring a similar system used at the Lily Dale Assembly, the Cassadaga Spiritualist Camp Meeting Association leased land to prospective homeowners to allow for semipermanence but to avoid the highly temporary nature of an assemblage of "shacks" or a "permanent tent community."[51]

Designed and laid out by the civil engineer Louis Redmond Ord in 1896, the original plan for Cassadaga was abandoned by the camp's board of directors in an attempt to open up the extensive public spaces of Ord's picturesque landscape for additional development. Following the rolling topography and establishing picturesque links with the lake, the town plan responded to the Spiritualist principles of connecting with nature. In particular, two of the National Spiritualist Association's principles adopted in Chicago in 1899 reflect this natural connection. The second tenet of the principles reads, "we believe that the phenomena of Nature, both physical and spiritual, are the expression of Infinite Intelligence," and the sixth principle states, "we affirm the moral responsibility of individuals, and that we make our own happiness or unhappiness as we obey or disobey Nature's physical and spiritual laws." In its current manifestation, the areas adjacent to the camp at Cassadaga have attracted an array of mediums, psychics, numerologists, palmistry practitioners, and tarot card readers.[52] Increasingly, the mediating zone of the campground is defined by the actions carried out within a particular location, but in many cases these activities also have connections to another place—whether imported by an itinerant preacher on his circuit or by a newly arrived numerologist.

The cleared camp in the forest has traditionally served as a site for the sacred negotiation of physical and spiritual laws. Within this uniquely American tradition of holding religious ceremonies in cleared camps within the open area of a stand of trees, the evolution from sacred *place*, an inherently spiritual zone as in the Greek's grove of Olympus, has evolved into a sacred *space* based on action.[53] Such action originates from outside a place and parallels many of Christianity's ritual practices of sanctifying spaces.

As spatial sanctity develops into hierarchical space and as cosmological time becomes liturgical time, the camp meeting offers a spatial and temporal antidote to religious and political orthodoxy. In the grove-camps, parishioners could experience a "private, sudden sanctity" in contrast to the "pre-ordained public progress" of conventionally mandated religious and social procedure. For the orthodox establishment, the first indication of this potential "spatial anarchy" was the itinerant preacher who, according to an anonymous writer in correspondence titled "The Wonderful Wandering Spirit," perambulates and "acts the busybody, is here and there and everywhere and above all things hates rules and good order, or bounds and limits."[54] This Wandering Spirit is soon supplanted by the Tin Can Tourists in their peripatetic movement across the Southeast and the nation.

Very early in the development of the camp meeting, the architect Benjamin Latrobe alluded to the relatively unstructured space of the campground and its potential for societal freedom in a collapsing of the sacred and the profane into one place, as a place-event. On August 6, 1809, Latrobe reluctantly traveled to a camp meeting held a few miles outside of Georgetown, Virginia. Although disparaging the meeting for its "incantations," Latrobe carried out a series of sketches that carefully interpreted the layout, organization, and spatial relationships of the camp meeting. The diagram that Latrobe made in his sketchbook presents one of the earliest graphic records of this phenomenon, which had become exceedingly common at the turn of the century. The semicircular form, with the stage at its center point, relates closely to the convex landscape and shows the regimented organization of classes and sexes. As indicated in Latrobe's precise sectional drawings along north-south and east-west axes, African Americans occupied tents at the lowest point along the stream, and men and women were seated uphill on either side of the axis formed by the minister's stage. In spite of this formalized separation, Latrobe found the experience of the place to be a wholly diverse public activity with "scattered inhabitants [in] a night scene of the illumination of the woods, the novelty of a camp . . . , the dancing and the singing, and the pleasure of the crowd, so tempting."[55] This unstructured and "undifferentiated"[56] space of the campground occurs between the sacred and the profane and allows the behavioral and social freedom expected by the parishioners and their spiritual advisers.

Although the turn of the century does not mark the end of the camp meeting, the transformation from spiritual to secular accommodates a new set of participants—the vacationing public. This change is helped along by many factors including the rise of autotouring, the growth of the vacationing middle class, and the World War economies. The semicircular form of the original camp meetings, with their paradoxically undifferentiated organization, continues as a religious basis for leisure community planning. The incompleteness of this circular form recalls the spectacle of the camp meeting and allows for the open program of temporary dwelling and recreational or natural experience. Rudolph Schindler adapted the half-circle form in the A. E. Rose Beach Colony project, an unbuilt experimentation with collective housing in 1937. Schindler synthesized the flexibility of beach cottages that could be moved and altered with the rigid definition of a communal space. In the design, a metaphysical openness to ocean and sky substitutes for the freedom of Latrobe's "tempting" woods. Schindler provided a collective but open framework for the Wandering Spirit of the individual tourist to live and play in a new social medium.

This form and its increasingly paradoxical implications for public and private recreation are further concretized in the New Urbanist experiment of Seaside, Florida. The new town of Seaside has been planned and developed by architects Andres Duany and Elizabeth Plater-Zyberk with developer Robert Davis. Since the development's beginnings in the late 1970s, popular literature, a series of photographic and textual monographs, and a special issue of the architectural journal *ANY* have both lauded and critiqued Seaside's ties to New Urbanism.[57] Seaside as "camptown" draws from the semipermanent settlements of the camp meetings at Martha's Vineyard and Chautauqua towns, and its plan registers Renaissance interpretations of Vitruvius's *Ten Books on Architecture* for ideal city plans of the Enlightenment.[58] Critics have also noted the potentially radical and innovative nature of this collusion of the classical treatise and the impermanent camp—contending that architect Duany "might have given birth to a gentle, friendly monster, a leviathan . . . relieving [Seaside's] determination to be a holiday camp for time-sharing puritans."[59] The amicable nature of the New Urbanist schema is debatable, but its influence on subsequent conurbations is undeniable. At the foundational core is the initial clearing for a camp space, one that has more recently been adopted as the image of the secure and livable postmodern city.

In addition to this confluence of classical form and the more informal campground, Seaside is a synthesis of Chautauquan ideals and the exigencies of late twentieth-century real estate ventures. The Chautauqua movement at its outset in the 1870s had already begun to structure the camp meeting as a more regimented event and permanent construction. Educated at Antioch College in the 1960s, the developer Robert Davis founded the Seaside Institute, which was modeled on the educational, pedagogical apparatus of Chautauqua. The second iteration of the Chautauqua movement was sited thirty miles to the north of Seaside in De Funiak Springs, Florida.[60] Located tangential to the spring-fed sinkhole called Round Lake, the settlement at De Funiak had begun similarly when Colonel W. D. Chipley camped along the lake with fellow surveyors who were laying out the route for the railroad line from Jacksonville to Pensacola.[61] The original grounds of the De Funiak Chautauqua included a public park area, a large auditorium, a hotel, and a series of residences.[62] In the 1930s, the Works Progress Administration restored the remaining buildings, but the annual six-week educational and entertainment sessions were no longer held.

While Chautauquas in De Funiak Springs and elsewhere have closed, Seaside has grown from its ambiguously socialist and utopic roots into a commodified economy of fashion, framed as a permanent campground for the wealthy. The urban code of this scheme relegates "trucks, boats, campers, and trailers" to the "rear yards only" while exempting the stylish "air-stream type" trailers. Seaside's design ultimately reflects a revernacularization of the classical, or the monumentalization of the vernacular if we reframe Colquhoun's concept of the "vernacular classical" as a post-industrial typology of urban form.[63] The town's regulation, through its codes, denies the possibility of a vernacular freedom that exists even in the most manufactured mobile home communities.

Clearing for Housing: Forming a Mobile Housing Strategy

> 200,000 trailers will swarm the roads this spring. Whether they betoken a New Way of Life or a plague of locusts is something that makers, taxpayers, hotelkeepers, and lawmakers are quietly disputing.
>
> —ROGER BABSON (1936)

Babson's apocalyptic declaration signaled the immediate need in the 1930s to address the relation between the mobile unit and the campground. The impermanence of the resulting assemblages of mobile housing would have substantial implications for the sense of American place in the latter half of the twentieth century. The story of this transformation in housing began with the experiments in production carried out by such visionaries as Corwin Willson. So permeating was this call to house the "plague of locusts" and the vacationing masses that architects of the time would find themselves trying their hand at industrial design and mobile home park planning. Even intensively place-conscious architects such as Frank Lloyd Wright and Rudolph Schindler found themselves caught up in the fever of mobilization.

Early recognition of the revolutionary dialogue between industry and inhabitant occurred in the field of architecture with Corwin Willson's publication of drawings for his "Mobile House" in *Architectural Record* in 1936. Willson's design reflects the migration of modern industrial trailer design into the realm of architectural discourse. In his detailed drawings, including plans and longitudinal and transverse sections, Willson proposed a sleeping loft above the compact 8' x 17'6" main floor. The design also called for lightweight molded plastic sections as the primary construction material. The trailer's interior spaces hold an uncanny volume and seemingly impossible program of occupation. In 1942, Schindler designed the prototype of a travel trailer for the George S. Gordon Sturdy Built Trailer Corporation. Given his background in such collective experiments as his own house on Kings Road, the A. E. Rose Beach Colony, and the universalist Schindler Shelters, the architect's unsuccessful mass production of the trailer is surprising in the design sense, but not unexpected in the architect's inability to slot into the industry, with its market-driven production. At the contrasting territorial scale, Wright's work at Arizona's Ocotillo Camp and the next iteration at nearby Taliesin West allowed for an application of his organic design process within the camping form.[64]

Manufactured trailers offer isolation and degrees of privacy within a relatively inflexible structure and form. The realities of occupying this "provisional frame" have been characterized as the "exigencies of the trailer's limited but oddly independent space."[65] With manufactured housing in general and camps in particular, it is necessary to differentiate the terms "trailer" and "mobile home." Put simply, trailers can be towed behind standard vehicles while mobile homes are delivered to a site typically by commercial trucks. As a result, trailers retain a degree of mobility that, ironically, "mobile" homes lack. Trailer and mobile home do share the common attribute that, as American dwellings, "their campiness is oddly resistant to nostalgic idealization."[66]

In both individual trailers and trailer parks as a whole, mobility and place sometimes find an uneasy synthesis. In some cases, the trailer park condition occurs between nomadism and land ownership as a confluence of impermanence and placefulness.[67] The question that arises is whether a sense of place requires permanence for its cultural and architectural value. For the architect Charles Moore, trailer parks actually demonstrate "mobility in the service of a sense of place."[68] Moore's concept gives a primacy to place in its relation to transient experience rather than creating place from within or with action defined by the mobility-permanence debate. Trailer parks themselves offer owners the possibility for constructing an individualized identity and a localized sense of place, however fleeting it may be.

The evolution of the impermanent and mobile dwelling forms an alternative historical thread of American domestic architecture. The early American wood-frame structures have been compared to the "legally mobile" dwelling of medieval land tenure. Along these lines, J. B. Jackson distinguishes the history of American ideas about home from those of dwelling and thus hypothesizes that the American house is a temporary construct that can gain, and sometimes does gain, permanence over time. Although housed in the semipermanent dwelling, contemporary notions of home have become fleeting and temporary. In such a model, mobility does not necessarily connote placelessness nor does it mean the demise of place. Instead, this newly emphasized tradition of mobility becomes a "new kind of home."[69] Reflecting the complex ideas tied up in the concept of camp, both resonance and paradox are found in the term "mobile home."

With the rise in popularity and accessibility of mobile home living and the increased production capacity following World War II, architectural interest in the trailer returned as campgrounds and trailer parks became the subject of design studies in architecture magazines of the 1950s and 1960s. Typifying these articles was a contribution by Frank Fogarty titled "Trailer Parks: The Wheeled Suburbs."[70] After critiquing previous trailer parks as "camp[s] of depression and wartime migrants," Fogarty proposed suburban layout patterns as antidotes to unacceptable campground design. Along these lines, his recommendation emphasized an increased lot size ostensibly to meet Federal Housing Authority minimum standards of the time and to alleviate the density of trailer parks and courts. These postwar recommendations followed from earlier sociological studies of mobile homes. To understand the house trailer as a dwelling type, this research in the 1940s explored the relation between preceding social frameworks and mechanical antecedents. The former looked back at smaller living units, smaller families, separate housing for the aged, mobility, and leisure time; and the latter studied covered wagons, tents, freight trailers, cloth-top trailers, and automotive travel in general.[71] Speculating that trailers would not replace the stable and fixed dwelling unit, researcher Donald Cowgill concluded that Roger Babson's 1936 prediction of half America's population housed in trailers within twenty years was alarmist and that, because the trailerites of the time were primarily those with mobile jobs and retirees, trailers were used only out of the necessity, or possibility, of mobility for work or leisure.

But the optimism associated with the mobile home's synthesis of mechanical and social needs continued into the 1960s. During this time, researchers proposed mobile home living as a viable type of renewable housing that could slow community obsolescence.[72] Such hopefulness persisted with studies such as *Urban Land*'s 1967 article "The Evolution of the House Trailer," which began with a historical overview of the decreasing mobility of the mobile home and the growing societal acceptance of it as a type of permanent shelter. A problem, however, lay in the uneasy planning relation between mobile home communities and suburban development. Zoning regulations and problems of taxation limited the formal synthesis of suburban development plans with mobile home ventures. Whereas the original mobile home park plans influenced the development of postwar suburbia in the 1940s and 1950s, the translation of scaled-up suburban plans back into the mobile home park design in the 1960s and 1970s met with popular and governmental resistance—the point at which we remain today. This evolution of mobile home design also marked a final split with suburbia when in 1970 the Frank Lloyd Wright Foundation designed a mobile home for National Homes. The conversion of Wright's Usonian home to a completely mass-produced, siteless vessel was complete. In 1952, Frank Lloyd Wright had designed a 442-space trailer park with additional motel lodging near Phoenix, Arizona. Not unlike a scaled-down version of Broadacre

City, the trailer park followed a grid scheme with small green spaces at each corner and a centralized, U-shaped greenway defining a communal and commercial space. But the success of Wright's scheme relied on a coordination of dwelling and urban plan—a negotiation that is not always cultivated today.

Subsequent architectural projects have adopted the trailer, or mobile home unit, as a fixed and modular unit of construction. Although influencing architectural discourse and later built projects, this speculative work has remained unbuilt and suggests how the avant-garde's camping excursions lack pragmatic experimentation—ground already covered, with varying success, by populist and vernacular planners. Wright's camps of horizontality, including the Phoenix trailer park and even Broadacre City, follow on his reading of the everyday, democratically inspired Jeffersonian landscape. Architect Paul Rudolph similarly adapts trailer culture to skyscraper development in a vertical "camp" for the Graphic Arts Center in Lower Manhattan. As the main component of a megastructure for the Lithographers of America Union, Rudolph proposed a trailer tower of between forty and fifty stories. Taken as an inevitable outcome of the economies of manufactured production, this construction synthesized the industrial with the architectural and the mobile with the fixed. Rudolph also sought to translate an American vernacular construct into the verticality of the high-rise condition. The trailer's module becomes the "twentieth-century brick."[73] Paul Rudolph had lived and worked in Florida for many years before settling in New York, and his observations of the local adaptations of trailer and mobile home culture are evident, not only in his Trailer Tower proposal but also in the modules and flexibility of such projects as the Beach Ball House and the Cocoon House.

The trailer becomes the "true vernacular" architecture, which "solves problems much better than architects do" and allows for everyday invention in the dwelling.[74] In Rudolph's Trailer Tower, a city within a city for Manhattan, the vernacular typology of trailer and skyscraper is pushed to their economic and structural limits. Each eleven-ton prefabricated trailer unit would be hung from 3"-diameter steel cables encased in concrete and attached to cantilevered "sky hooks" that extend from vertical hollow tubes in which circulation and mechanical components are enclosed. Echoing terminology used by the British group Archigram, Rudolph calls the units "capsules," with 3 1/2"-thick corrugated steel walls and standard 12 x 60 x 8' dimensions that fold out to 24'-wide floors and roof terraces once in place. The roof of the lower mobile house, or "truck van," becomes the terrace for the higher unit. Rudolph contrasts the lightness of his idea with that of the concrete heaviness of Moshe Safdie's Habitat project for the 1967 Montreal Expo, and he differentiates his trailer tower from Archigram's Plug-In City in the necessary permanence achieved with the units once they have been "plugged in." Here, Rudolph responds to the typical fixedness of mobile homes: "Once they are there then they are there. The portability is a misnomer to a degree."[75]

Clearing the Avant-garde City: Remaking the Gypsy Camp

instant villages . . . (camping scene not included)
—DAVID GREENE (1966)

Between 1962 and 1964, Peter Cook of Archigram designed the Plug-In City, a project reflecting one technocratic interpretation of what a campground could be. As an avant-garde project par excellence, the Plug-In City was proposed at a megastructural scale, with a hierarchy based on the degree of impermanence. Cook described the city as being "set up by applying a large scale network-structure, containing access ways and essential services, to any terrain. Into this network are placed units which cater for all needs. These units are planned for obsolescence."[76] The Living City exhibition, at London's Institute of Contemporary Arts in the summer of 1963, set the theoretical and methodological background for Archigram's work with similar urban environments. Much like the programmatic openness found in camping facilities, the Living City was not a blueprint for urban space. The city was instead an "organism housing man" with its purpose "to capture a mood, a climate of opinion, to examine the phenomena of city life."[77] Members of Archigram followed up on this initial work with projects for a Walking City in 1964 and, two years later, a Drive-In Housing unit with the premise that the individual could make a home wherever the container came to rest or was parked. The Free Time Node Trailer Cage in 1967, much like Paul Rudolph's proposals for Manhattan, continued these experiments and appropriated the mass-produced trailer for the singular and often extraordinary practices of vacationers at seaside resorts, participants in festivals, and workers in remote areas.

The influence of camps and camping on Archigram's work can also be found in David Greene's essay "Gardener's Notebook." This project included another kind of camping node, identified as the Locally Available World Unseen Networks (LAWUN). The LAWUN, like the travel trailer idealistically deployed in a more natural environment, drew from Leo Marx's *Machine in the Garden*, particularly through Greene's explication of the Bot as a machine "transient in the landscape."[78] Greene proposed a landscape of proprietary Rokplugs and Logplugs that provided a dispersed, invisible infrastructure within the garden that was itself a remnant from the decay of both urban and suburban developments. These outlets, hidden in boulders and logs, allowed for "instant villages . . . (camping scene not included)" and were ironically offered as a solution to the camper's problem of energy sources for "mobile living support systems."[79]

Though conceived along different technical lines, the work of the Situationists influenced Archigram's projects for the city, particularly those that addressed the production of space.[80] If its satire of functionalist consumerism is disregarded, Michael Webb's "Sin Centre" of 1962 can be compared to models for New Babylon of 1958 in terms of its scale, flexibility, and conception in model form. We also find the influence of Johan Huizinga's *homo ludens* with both groups' interest in the role of "play" in urban situations. Two of Huizinga's primary theses are that "pure play" is a basis of civilization and that play, or the human ludic quality, is inextricably woven together with culture. The characteristics of play outlined by Huizinga generally circumscribe the activity of camping itself. For these avant-garde architects, play served as a flexible program for urban events: "The play-mood is labile in its very nature."[81] Play diverges from ordinary life both in terms of locality and duration. Like the festival, the place of play is often marginal and its time of occurrence is limited in actual duration but endures through a limitless memory and recollection. In this sense, play is "done at leisure, during free time." And the extra-ordinary aspect of play serves as an interlude between episodes of daily life: "It is . . . a stepping out of 'real' life into a temporary sphere of activity with a disposition all of its own."[82]

In the projects of the Situationists and Archigram, play is transformed from a marginalized activity subordinated to seriousness and the expedients of production to a central programmatic procedure for architectural design. Henri Lefebvre, who was influenced by Constant's *Pour une architecture de situation* (1953), also developed the idea of play as a vital feature of culture. For Lefebvre, the "old places of assembly" in the city have been abandoned by the event that he calls the *fête*. He advocates the discovery and construction of places appropriate to a renewed *fête* that is fundamentally linked to play.[83] This revitalization of the *fête* allows for the centrality of play that results in the privileging of *time* over the inflexible *space* of production. For both Archigram and Huizinga, this idea of free time is not simply leisure time but goes deeper to include a time of fluidity—apart from history's linear time, a folded time of coincidences and overlapping moments. At the "apogee of play," Lefebvre proposes an "ephemeral city, the perpetual oeuvre of the inhabitants, themselves mobile and mobilized for and by this *oeuvre*." In this idealized ephemeral city, Lefebvre notes that the centrality of play makes space obsolete and instead privileges inhabiting over habitat. The "art of living" is no longer reduced to its spatial manifestation.[84] As a flexible practice and an open field, camping also emphasizes the "place" of play over the "space" of leisure. Camp privileges the qualitative experience of "getting away from it all" over the quantitative features of the camp.[85]

In 1958, Constant produced the Model for a Gypsy Camp, which became the basis for his subsequent models of New Babylon. Constant's relationship with Giuseppe Pinot-Gallizio and the early developments of the Situationist movement influenced his production of the Gypsy Camp model. Pinot-Gallizio had worked as an advocate and political representative of the Gypsy population that visited the town of Alba and also organized the Experimental Laboratory of the International Movement for an Imaginist Bauhaus in Alba, where Constant was inspired to develop New Babylon. After becoming a member of the experimental laboratory, Constant attended conferences in Pinot-Gallizio's studio, located in a seventeenth-century monastery. And then in December 1956, Constant visited a Gypsy camp on Pinot-Gallizio's property along the Tamaro River. The Zingari Gypsies had been banished to the site after camping under the roof of the town's livestock market where "they lit fires, hung their tents from the pillars to protect or isolate themselves, [and] improvised shelters with the aid of boxes and planks left behind by the traders."[86] The town council had determined that the Gypsies did not clean up sufficiently and had banned them from Alba's public spaces. In the encampment along the Tamaro, Constant discovered the Gypsy town: "they'd closed off the space between some caravans with planks and petrol cans, they'd made an enclosure." This event inspired the initial model and the larger utopian project: "That was the day I conceived the scheme for a permanent encampment for the gypsies of Alba and that project is the origin of the series of maquettes of New Babylon. Of a New Babylon where, under one roof, with the aid of moveable elements, a shared residence is built; a temporary, constantly remodeled living area; a camp for nomads on a planetary scale."[87] Early in 1957 after preparing a psychogeographic program for Alba and designing a pavilion for the laboratory, Constant returned to Amsterdam and developed the Model for a Gypsy Camp—"the first mobile architecture of *urbanisme unitaire*."[88] The Gypsy Camp included movable dividing walls that could be manipulated by the nomadic inhabitants, and its overall design resembled a tentlike circular tensegrity structure. Lefebvre's "ephemeral city" had been realized as an architectural scheme.

Pinot-Gallizio's 1959 exhibition of work in Paris at the Galerie René Drouin reflected the Situationist vision, also inspired by Lefebvre, for a diverse urban festival. One purpose of Pinot-Gallizio's exhibition was to show that "free time, rather than being filled with banality . . . could be occupied in creating . . . massive architectural and urbanistic constructions."[89] With its enclosure of painted canvas walls and ceilings, the exhibition-cavern resembled a series of Bedouin or Gypsy tents, forming a labyrinthine camp of unfurled fabric. Later that year, Constant exhibited his work for New Babylon in Amsterdam, where the display of his "model precincts" was "inspired by unitary urbanism—the design of an experimental utopian city with changing zones for free play, whose nomadic inhabitants could collectively choose their own climate, sensory environment, organization of space."[90] In addition to the influence of Pinot-Gallizio's work with Gypsies and his camplike exhibitions, Constant's models of New Babylon arose out of his reading of the Lettrist method of *dérive* and the psychogeographic understanding of urban ambiance. For Constant, the correlation between the Gypsy Camp and New Babylon fused the vernacular and the avant-garde tendencies toward the chronic, and societally acceptable, ephemerality of play. Constant later reflected on the relation between camp and city: "If urban space were planned to meet the needs of a leisured society, these flights from the city would become unnecessary. Paradoxically enough, when townspeople trek en masse to the great outdoors, the difference between town and country disappears. A camping area is a form, however primitive, of a city."[91] Camping form elides differences of country and city, as *chora* and *polis*, to become a basis for a new kind of space for the city—granted one that is utopian in its premise.

To formulate his avant-garde city, Constant studied how Gypsies camp within the town's places—clearings of temporary refuge with their own labile modes of organization. Constant also noted the paradox of mass camping, which in effect mobilizes the city and reconstitutes its form elsewhere—overriding spatial differences of city to country and accommodating the idea of place through activities of leisure, recreation, and camping. Pushing these ideas to their theoretical limit, Constant proposed a camping milieu at a global scale, an infinitely extensible matrix, temporary in its flexibility rather than in its placement. Constant fashioned these ideas as maquettes and thus carried out the temporary fixity that the making of camp entails. "Clearing" opens up the possibility for the free play of making place anew.

PART II
MAKING

July 1991 (Athens, Greece)

I spent a summer documenting stones at the Athenian Agora for the American School of Classical Studies in Athens. Each weekend my wife and I toured the Greek countryside to see the ancient sites, relying on rented mopeds (papakia) and public transport. Most destinations were remote, but we were determined to maximize our time off from work, and overnight lodging was neither available nor affordable. So, we camped.

We moved from archaeological site to archeological site with our sleeping bags and toothbrushes. We made our camps based on location, shelter, and proximity to the next day's destination. We sometimes woke to the hum and suffocating exhaust of idling tour buses. At Sounion, we set up camp after dark and unwittingly made our campfire, had dinner, and slept along the edge of the site's parking area, indeterminate at night but clearly defined by day. At Naphlion we camped on the beach, and at Delphi we lodged on the side of the road.

As we made our minimal, improvised camps, we became, for the moment, Greek herms—personifying the ritual process of travel. We were witnesses to the sites, and at the same time we were roadside attractions, oddities of an alternative and impractical tourism. Our campsites were as accidental as they were planned. Travel for us was a hermetic gift, truly a "thing in the making"—not completely ad hoc but measuring a subjective landscape against the great Hellenistic legacy, a sublime historical presence made intimate through our method of occupying its imago mundi.

As made things, our camps were details within the archeological territories of the Greek countryside. Derived from Hermes as the god of boundaries, travelers, merchants, and communication, ancient Greek herms were piles of stones deposited by passersby often at crossroads or points of interest. Although not shrines, the camp marks a node (sometimes sacred in its transcendentalism) within the larger network of travel and its ritually charged landscape. The making of herms through time reflects the archaeological process of excavating. Piling up stones, building layer upon layer, and making and remaking camp articulates a fragile accretion of material and memory.

On our return to North America, we recognized in our Greek camps a happy coincidence of presence and absence, of "real" time and historical time. Although our campsites were adjacent to notable excavations, we realized that we had been camping in a larger archaeological site—a place that was continually being made and remade as it was occupied and as it was simultaneously being dug up. Like the Greek god Hermes, making camp is a relay between time and place and between culture and memory.

Chapter 4

MAKING CAMP

Architecture does not make things it sees but watches how things are made.
—QUATREMÈRE DE QUINCY

Watching How the Vernacular Is Made

The vernacular in architecture is a process. As a vernacular construction, the camp is then a practice of making and thinking. The link between vernacular and process extends back to Dante Alighieri's treatment of vernacular language in *De vulgari eloquentia.* In his work, Dante compares the vernacular to a living organism, specifically the human body. Dante uses a man's development to maturity as a metaphor to describe the process of the vernacular's growth: "Nor should what I have just said seem more strange than to see a man grown to maturity when we have not witnessed his growing."[1] According to this model, the vernacular is a being in the process of becoming.[2] Such a dynamic proceeds along an "unstable ground" of difference and contradicts vernacular's commonly held meaning of similitude and uncomplicated locality.[3] Linking language and humanity, Dante proposed that the "illustrious vernacular" be composed of suitable fragments from existing and nascent languages.[4] Though Dante hoped that the vernacular language would become the standard language of citizens, poets, and courts alike, the initial formulation of the language began from the difference and the dissimilarity of dialects and regional influences. The organism of the vernacular however remained a "living system" that could adapt and evolve with change. Like camp, Dante's language system is not only flexible and open but also mobile. Dante personifies the new language as a “homeless stranger.” This itinerant formulation comes out of Dante's focus on the Italian poetry of aulic and courtly troubadours, who out of necessity are at home equally in domestic and royal spaces.[5]

Camping is a wandering vernacular for which the “whole world is a homeland.”[6] The history of camping is carried out along roadsides, in backyards, and in established campgrounds—all defined by populist methods of place-making. The vernacular process is paradoxical in its negotiation of mobility, place, and permanence. The vernacular has been identified with transient communities, such as mining, shipping, and military settlements. These vernacular camps respond and adapt to local conditions, but at the same time are constructed with “materials and techniques imported from elsewhere.”[7] The indigenous and local characteristics of the vernacular are complicated by the influence from “elsewhere,” or, more precisely, “another place.” Delving deeper into vernacular's origins reflects this connotative complexity of a specific elsewhere. In Latin, *verna* refers to a slave born abroad—that is, a slave born within the master's home but away from the *verna*'s native land. This particular distinction, while alluding to an etymological disconnect somewhere along the term's linguistic evolution, also reflects two important themes: an exteriority incorporated from within and a temporary presence in the process of developing some degree of permanence. It is with camps, campsites, and camptowns that these paradoxes become integral, even foundational. Paradoxical conditions arise from the mobile fixity, unstable permanence, and chronic itinerancy of camps. Michel Butor captures this idea in his description of the unsolidified permanence and hardened ephemerality of the vernacular camp constructions that became Istanbul: “An encampment that has settled, but without solidifying completely; huts and shanties that have been enlarged and improved, that have been made comfortable, but without ever losing their ephemeral feeling. Turkish Istanbul . . . is truly the expression of an empire that collapsed on itself as soon as it stopped growing. In the great bazaars awning had turned into roof.”[8] The encampment, though enlarged and improved, does not lose its transient quality, even within a growing permanency. Each of the camps investigated here maintains a degree of permanency that is complicated by origins in the ephemeral.

At the level of detail, vernacular as process and paradox also relates to the improvisation of assembling. In camping, this improvisation of the assemblage is not an ad hocism of socially plural or formally composed fragments.[9] The emphasis on process, necessitated in camp situations and their study, disallows the notion of the ad hoc as a simple expression of social pluralism or as a formal composition of fragments. The ad hoc of camping vernacular, while self-regulated to a great extent, is inextricably tied to its local operations, which is an absolute manifested locally. As a Deleuzian "local absolute," the ad hoc then is not an overlay, but an internally generated event that is "grounded" in the place. The imperfect repetition of the vernacular process provides a stability, as ground and as way of making, that does not result exclusively from individual expression. The success of each operation relates to previous successes in a partially empirical mode. Improvisation is half empiricism and half systematization constrained by rules, code, or material limitations.[10]

As researchers, geographers, architects, and place-makers, the ad hoc offers for us not a license of pluralism or a promise of the spontaneous creation, but instead a method of looking at the way things are made rather than just at what is made.[11] It is possible that the lessons learned in the study of the vernacular through camps and camping can be folded back into a process of design that avoids mimesis and achieves a poetic relation between the built environment and place. Improvisation, in this sense, relates to bricolage. Casey compares the "transitory nomadic camp" to what he calls the bricoleur's home laboratory, which is "set up with materials ready at hand in a casually arranged workplace" that lacks the fortification of walls.[12] Vernacular is that which is made, and it is the process of making that gives it meaning. Camp, as a vernacular construction, is bricolaged improvisation.

As Dante's "living organism," the vernacular manifests itself not merely as a building[13] but as a building process. The question becomes one of *how* rather than *what*. The vernacular built environment occurs as a dialogue and as a set of operations between the detail and territory—an operational condition in which the building process must negotiate this scalar divide. With its materials and techniques imported from "elsewhere," the vernacular as process takes us from the highly localized architectural detail to the extensive cultural exchange created by oftentimes external cultural, economic, and social forces. The buildings and constructions created are artifacts of this process. With such complexity and often paradoxical interconnectedness, this artifact, as camp, becomes a "specific elsewhere" reflecting both local and global constraints and possibilities.

Watching How Camps Are Made in Place

The objectives of each camp also register attempts at a synthesis of a local-regional interaction.[14] Ruskin is a socialistic utopian settlement, Manila Village is a fishing platform community, Gibsonton is a camptown for carnival performers, Braden Castle Park is a permanent tourist camp, and Slab City is a camp for retirees and the homeless. The making of these camps moves between territory and detail in their complex constructions within these places. From the review of central Florida's camps, the making of Ruskin ranged from the 12,000 acres of its timber plantation to the 160 acres dedicated to its college to the nominal acreage take up by its original shell midden and camp. Braden Castle Park's 34 acres of land divide into 900-square-foot properties, and Slab City's sprawling 640-acre territory can be contracted to the individual study of its particular "slab" constructions, providing a series of 600-square-foot platform sites.

Details of how the camps are made are also found in the connections and assemblies of the dwellings. Ruskin's traditional pine slab and board-and-batten cabins contrast with Gibsonton's modified mobile homes and trailers with manufactured concrete-block foundations. Slab City's dwellings range from assemblages of discarded military ordnance to Class A motor homes parked in communal configurations. Manila Village's pole construction and partially floating walkways reflect the density of Braden Castle's bungalow constructions scaled to the autocampers and travel trailers they had used as tourists. In Florida's municipal parks, it is the connections between tent and automobile that define the spaces of those early camps.

Making Camp as Accretion

The making of camp proceeds through accretion and association, existing in and external to the particular place. Some camps are overlays onto previous uses of the site. In Florida, the Ruskin College campus used the buildings and grounds of a turpentine camp for its formulation. Gibsonton's

Walker Evans, Recreational vehicle with striped canopy, 1941. Gelatin silver. 5 3/4" x 8 11/16". The J. Paul Getty Museum, Los Angeles.

site along the Alafia River was originally a fish camp. One of Florida's first municipal campgrounds began as a public recreational area known as De Soto Park, where the Tin Can Tourists first met in 1919. Braden Castle Park utilized the homesite of the Braden family sugar plantation. And Slab City was sited on the residual components and layout of the temporary naval training facility Camp Dunlap.

The camps also draw from external sources for how they are made. In its self-regulation and cost-free living, Slab City relates to the Long Term Visitor Areas of the Southwest. Braden Castle Park shares attributes of new town planning of the 1920s and 1930s as well as other tourist camps in Florida. Gibsonton draws from the carnival midways where its inhabitants perform each season. Manila Village incorporates the making of Filipino boats and stilt structures into its construction. Ruskin's town layout resonates with earlier Ruskin colonies in the United States.

Based on its construction, form, and layout, each camp also elicits theoretical and conceptual associations. Ruskin's college campus and newer buildings relate to John Ruskin's work with Oxford University and the Guild of St. George as well as to his written work. Manila Village resonates with the floating houseboats of Delta and Atchafalaya "swampers" and with the historically distant though apposite construction of the Achaeans' city of beached, or "landed," ships. As the original "midway," the Midway Plaisance at the World's Columbian Exposition of 1893, which itself included an array of exhibits staged as camps, provides an archetypal and procedural frame for the town of Gibsonton. In a folding of time, Tampa's Municipal Campground becomes a Roman Campus Martius with its public space of low-lying, flood-prone river frontage. With its pragmatic bylaws and spatial efficiency combined with the sublime environment of its romantic ruin, Braden Castle Park contrasts with New Urbanist planned developments that have their origins in Chautauquan experiments in community but have become seasonal tourist settlements driven by nostalgia and displaced sentimentality. Finally, Slab City resonates with the hierarchies and formality of the Roman *castrum* while also mirroring elements of the Autonomous Zones of yearly event-cities such as Burning Man.

Making Time for Camp

The making of these camps occurs across historical time, but at the same time each camp is an intensively present version of place-making. The time of camps is pulled to the present, and historical development informs the current complexities of the camp's construction of place. In camps, time is often controverted: pasts are, and presents were. Such coincidence is not merely an inversion but a porosity of time within the relatively thin historical layering of the accreted campsite, best illustrated in the dense but short history of European occupation of camps in the Florida landscape.[15] Camping practice, with its relation to method, lends itself to these coincident times.[16] The time of camping in concert with its making is ultimately a layered time, similar to the architecture of waiting that Paul Virilio found in the World War II bunkers of France's "Atlantic Wall." Carrying out his "solely archaeological" objective to hunt these gray forms until they would transmit . . . a part of their mystery," Virilio arrived at an archaeological time characterized by "rupture in the apprehension of the real."[17] Camping occurs in "real time," susceptible to the specific conditions of its places and the making of its campsites.

Although connected to archaeological constructions, this time is not simply stratified. The layering is in fact often broken and folded. In archaeology, stratigraphic disruption is the norm, and contiguity is anomalous. Archaeological time is consequently nonlinear and can be understood as a series of broken cycles. Such is the cyclic time of Gibsonton's carnival worker population and Slab City's "snowbird" inhabitants. Though a profane secularization of native culture, the material of Ruskin's shell midden remains in the surfacing of its streets. The made time of camping is also prospective, so that Virilio's taxonomy of the bunkers is based on the different "manifestations of time" found at each site. Braden Castle's ruins are the focus of the community's public grounds; and the slabs of Slab City, as the most permanent material of the site's previous use, provide foundations for exceedingly temporary shelters. Even Manila Village's support posts remain, though encrusted and wracked by tidal fluctuations—echoed in the detached floating houseboats of Barataria Bay and the Atchafalaya Swamp as well as the more massive supports of the Delta's oil derricks. Archaeological time alone cannot complete our understanding of these camping situations—such that the activity of "watching how things are made" then requires us to reforge tools to circumscribe our own camping space.

Living Formatives and the Makeshift Camp

> . . . after the manner of . . . of the nature of . . . pertaining to . . . of . . .
> —*Oxford English Dictionary* (2nd ed.)

Living formatives frame and expand on the significance for place-making held within each camping practice. Forged around a specific analytic-poetic term, each word with the suffix "-ic" is a provisionally assigned operator that arises from field research and serves as a "makeshift" ground for the preliminary study and analysis of the place.[18] As descriptive terms and as exploratory terminology, these formatives are not framing devices because the campsites, as places and living case studies, outline their own narratives and historical spaces. So the living formatives are neither applied normatively nor are they passively reinforced by the particular place's attributes. The formatives instead work in conjunction with the specific places to comprehend the broader significance of making camp and participating in the camping process.[19]

Each camp's analytic-poetic term derives from its making. As an example, the platforms of Manila Village are "asymptotic" in their relation horizontally to the tidal water surface and tangentially to the territorial waters of the United States. If this method of understanding place also entails the process of navigating the resonances and layerings of each camp(site), then the other aspect is bound up in the question of how to analyze and understand the relevance of the made "thing." This idea returns to the formulation of "making" as a noun. Each of the "-ic" terms qualifies and modifies the stability of the camp as a completed, or seemingly finished, site of construction.

"Taxonomic" relates the complexity and difficulty of classifying the proliferation of campsites, particularly in the case of Florida's varied places. Such an attempt at classification necessitates a consideration of differences rather than a categorization of simple equivalencies. These differences may be in kind or in degree. In Raoul Bunschoten's project *The Skin of the Earth*, an inventory of domestic fragments becomes "a taxonomy of degrees of familiarity or alienation."[20] Thus, the hybridized fragments, which Bunschoten calls "domestic *metafora*," can be read through their taxonomic difference, registered in each component's defamiliarizing tendency. The taxonomy ultimately aids in negotiating the project's manipulation of the ground and its being recast in a domestic role.

In its use here, taxonomy becomes a labile grammar based on difference. This framework approaches the syntactic variability and improvisation found in the vernacular languages by Dante Alighieri and much later by William Labov in both figures' work with "nonstandard" language.[21] With this emphasis on the often subjective arrangement of words, as in the improvised constructions of camp, camping taxonomy can be differentiated from the methodologies espoused by structural linguists of the 1940s and 1950s and criticized by Noam Chomsky for their disregard of the perception of language. In the unique situation of camping, perception and detachment with respect to other communities and the natural world define the camper's rhetorical practice of place-making. Thus, rather than Michel Foucault's *taxinomia* as the science of articulation and classification found in *The Order of Things*, the use of "taxonomic" is more closely tied to its root in the *taxis* defined architecturally by Vitruvius. Here, *taxis* and thus taxonomy do not simply result in an order but provide an order*ing*. Such ordering occurs in a modular layout rather than within the confinement of a grid.[22] "Taxonomic" thus works in three ways: how the permutations of camps within the Florida context refer to a generic Floridian camp and at the same time refer to a specificity of place and intention; how, taken as a whole, the Florida camps constitute a taxonomic grammar of regional camping practice; and, as adaptations of existing sites with their own system of ordering, how ordering principles can be found in each camp along with an inherent connection in each settlement between territory and detail, typically through proportional relationships.

"Eutopic" refers to the qualities of a place of ideal happiness and good order. The term arises from the tourism and optimism of the Tin Can Tourists in their pursuit of a Floridian eutopia. Braden Castle Park, in its founding principles, bylaws, and architecture, resonates with the tourist's search for the ideal place to visit and ultimately to settle. "Eutopia" diverges from Sir Thomas More's imaginary island of "Utopia" and its theoretically perfected legal, social, and political system. It is the impossibility of utopia's "no place" that the pragmatics of eutopia necessarily circumvent. The eutopia of Braden Castle Park does however include the indefinite remoteness of utopia's ostensibly uncharted places. The park forms a veritable island in its informal status as a city within a city. And the park's laws also reflect the expectation of social improvement found in More's utopia.

"Parasitic" corresponds to that which maintains and transforms. Typically, parasites are understood as unwanted guests within or attached to yielding and unknowing hosts. In other cases, however, the parasite maintains and transforms the *relation* between host and guest.[23] This concept yields three significant modes for the parasitic conditions found in camps: parasite as excluded middle, parasite as both static term and operator in a system of relations, and the time of the parasite as duration. Giant's Camp and the town of Gibsonton are reviewed for their parasitic qualities in each of these frameworks. The practice of camping often relies on this parasitic relation to places, host cities, and environments. Giant's Camp, in particular, over a period of time serves as agent of both preservation and change for the larger camp community of Gibsonton. In these cases, where camps reach a point between the temporary and the permanent, the "parasitic" reflects the importance of considering time and place, rather than exclusively space, in the study of natural and built environments.

"Heterotopic" questions the connection between Foucault's heterotopia and the construction of camp space. Although it is not a heterotopia as such, Slab City shares characteristics with Foucault's heterotopic colony and the paradoxical place of the metaphorical ship. The "heterotopic" thus includes both dis-placement and mobility of place. Slab City can be read as a colony of domestic exile in which the camping public defines its own rules for constructing and living in the built environment. Such self-regulation does not imply the complete dissolution of codes of conduct and building, but results in a consciousness of "difference" and "otherness" that is further complicated by an internal differentiation of class.[24]

"Asymptotic" refers to a proximity that retains a distance. Such a paradoxical situation can be described colloquially as "almost touching" or "closing the distance." The platform communities of Manila Village maintain this propinquity and remoteness along with attachment and detachment. The "asymptotic" characterizes two main relations within the constructions of Manila Village—the territorial tangencies of the community with the ambiguous political boundaries of the state and, at the scale of the detail, the relation between the fluctuating surface of the tidal waters of Barataria Bay and the horizontal surfaces of the platform structures. In both cases, the "asymptotic space" can be understood as a variable space-in-between. More recently, the spaces of postdisaster urban environments, like New Orleans, have become the sites for a new asymptotic city of camping, with significant implications for how we might continue to make urban places and more broadly how we will dwell.

Working at the scale of both detail and territory, the "asymptotic" suggests a potential combination of the diagrammatic and the tectonic for us to understand how constructions might be made. Like the "asymptotic," the diagram serves as the intermediate generator between processes of thought and making. Consequently, the diagram of the asymptotic becomes an "icon of intelligible relations" among earth, land, water, materiality, and territory. This iconic presence points *toward* meaning rather than proscribing a closed meaning—the open-ended and sometimes symbolic potential of camping practices.[25]

Chapter 5

CAMP(SITE)

Florida's Vernacular Places of Mobility and Temporality

Camp Space

Florida's peninsular condition, at the southern edge of the Deep South, has historically attracted disenfranchised and otherwise itinerant groups seeking space to create settlements to develop alternative social structures and to practice marginalized philosophies. At the same time, the Florida region has accommodated the passing aspirations of tourists. Both groups have in some way appropriated or transformed the campground typology. The idea of "camp space" moves from its materiality as a place of temporary dwelling to its more intangible aspects as a mentality and a marginalized aesthetic. Within this range of investigation, "camp space" becomes a thread that conceptually links four nodes—Ruskin, Masaryktown, Gibsonton, and the places associated with the Tin Can Tourists, geographically connected by a section of Highway 41 near Tampa, Florida.[1] The marginality and "southernness" of these campsites characterize their significance locally and regionally and their qualities of geographic location, climate, economy, and culture. The impermanence of camp, as an idea and a place, also suggests more open-ended linkages to places outside Florida and to related practices of place-making.

The negotiation of the spatial paradoxes found in camps redefines the Florida architectural vernacular. Previously understood as style, definitive typology, or strictly regionalist production, Florida's vernacular built environment cannot be fixed as form or image and instead must negotiate the paradoxes of camp space. John Ruskin in the subtropics, Ferris wheels in front yards, and Thomas Masaryk's supporters in logging camps. Philosophers, photographers, carnival performers, revolutionaries, tourists, and migrant workers all camp out in the state's humid, sometimes inhospitable, bristling landscape. The Works Progress Administration's (WPA's) *Guide to the Southernmost State* attempted to stitch together this vernacular patchwork of burgeoning ports, interior enclaves, and already artificial natures. The guide's collaged format paralleled Florida's distinctive confluences of foreign autochthony. Previous guides to camping in Florida had emphasized either the logistic realities of travel to the state or satisfied the armchair curiosities of those unwilling or unable to reach its extreme "southerness" in the 1920s and 1930s.[2] The WPA guide provides for us an unusually poetic manual of the region's places and oral histories.[3] The ironies of the Deep South are redoubled, and we find humor, pioneer bravado, and a degree of campiness in its stories and in its cautionary tale to tourists:

> *Caution to Tourists:* Do not enter bushes at sides of highway in rural districts; snakes and redbugs usually infest such places. Do not eat tung nuts; they are poisonous. Do not eat green pecans; in the immature stages the skins have a white film containing arsenic.[4]

This warning is a reminder to the unwitting Florida visitor, tourist and nonnative resident alike, that whether you like it or not you are really camping within this "southernmost" experience.[5] Implicit in this warning statement, and throughout the guide's many tales, is the idea that the region's interactions and inversions of the temporary and the permanent necessitate the development of an alternative process for occupying its shifting landscape.[6] Forgoing its didacticism and nostalgia, Florida camping is a point of provocation, a procedural model, and an imagery base for discussing contemporary questions of how to understand and occupy proliferating spaces of itinerancy.

The area of focus in central and south Florida shares a subtropical climate and exhibits minimal changes of elevation in its topographic landscape. Beyond these similarities and the sites' linkage through Highway 41, the region's demographics, history, settlement and land-use patterns, and

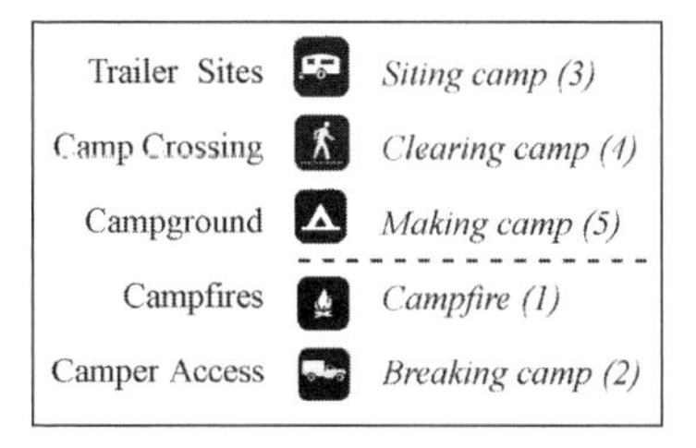

Camp legend: siting, clearing, making, campfire, and breaking.

stages of development vary considerably, even with minor shifts in location. Consequently, this region is a network of unique locations and events that together formulate common themes of itinerancy, tourism, and utopianism. This indeterminacy and diversity actually contributes to the significant possibilities of what I have termed "camp space"—a working method for understanding and occupying fluid and dispersed places. This chapter adapts the camping sequence for the explication of the cases and to outline what "camp space" is and what a camp practice might be through the chapter's taxonomy of places.

The legend above associates camp icons with stages of camping. On a literal level, the sequence of camping is repeated as a cyclical process: we look for a suitable site, we clear debris from that site, we set up camp, we build a fire and tell stories, and after covering fading embers we break camp to repeat the procedure. In early Florida, campfires were fueled by the resin-saturated pine called "lighter wood." These fires provided the typically communal focus at the end of the day: "Sitting around the blazing pine logs that night, the time passed quickly while talking of hunting, fishing, and sailing, and it was eleven o'clock when we turned in."[7] At a practical level as a series of events, camping is a circumstantial and highly conditional activity. Camping is thus repetition with difference; and in the overall sequence, events can overlap or occur simultaneously.

Within camp, this differential quality is also found in the recounting of stories, legends, and myths around the campfire. The oral retelling of legends modifies and blurs distinctions between events. In Florida, this oral history includes stories found and retold by Stetson Kennedy as the director of the Florida Writers' Project unit of the WPA between 1937 and 1942.[8] In one section, Kennedy recounts the harsh life of turpentine camps in the stories and songs of African American laborers.[9] In her proposal for interviews in a west Florida turpentine camp, Zora Neale Hurston also defined areas of Florida based on their oral history and tradition of songs and folklore. In Area III, from Palatka to Florida's west Coast near Tampa, Hurston writes of the rich storytelling and singing traditions found in the region's camps: "The most robust and lusty songs of road and camp sprout in this area like corn in April." Hurston concludes the text with the summary appellation: "Florida, the inner melting pot of the great melting pot, America."[10] The variations along with particularities of circumstance found by Hurston within these camps reflect one aspect of the paradoxes of "camp space."

The cyclic nature of camping is also susceptible to interruption. Disturbances in Florida range from natural occurrences such as wild animals, rain, wind, cold, or uncontrolled fire to social infractions such as irate landowners or disagreements among campers, to technical failures such as the breakdown of equipment mired in the swampy grounds. The dashed line in the legend above represents the break in the camping process—a procedure that is both sequential and cyclical. With this characteristic in mind, I have reworked the sequence of camping to begin "around the campfire." The campfire is an originary and centralizing feature from which the camping practice proceeds as a series of four events and from which the practice's operational modes radiate. At a metaphorical level, this rupture within camping's typical progression serves to "localize" this chapter's own event—I am, in turn, retelling the story of Florida's camps.

The "legend" introduces and indexes not only literal and metaphorical considerations of camping but also illustrates two coexisting trajectories: the explanatory and the exploratory.[11] In mapping, the legend provides a key for reading the map; and in mythmaking, the legend provides possibilities for "opening up" and making spaces, as implied by the transmission of stories in *mythos*. Used here as methodologies, both mapping and mythmaking serve as means of communication that operate between the analytical and the synthetic. Analytically, aerial photography of the Florida peninsula has served both agricultural and promotional ends. Early aerial imagery from the 1930s and 1940s satisfied the cadastral objective of defining properties for agriculture. This imagery also now provides an important document for understanding the location and layout of Florida's early places, from turpentine camps to tourist campgrounds, many of which can be discerned only from the air.[12] As an enticement to tourists, the Florida Chamber of Commerce and Eastern Airlines contracted with the Aero-Graphic Corporation in 1936 for three flights over the state: "To travel it by land, is to visit a tropical paradise. . . . To cover its thousands of mile shore lines [*sic*], . . . is to be one of nature's chosen few who are privileged to view so magnificent a spectacle."[13] On the ground, the legendary natural attributes of Florida are synthesized into the Indian mythology of the region supplemented by folklore of both African Americans and white settlers, known colloquially as "crackers."

The introductory legend above serves as a guide for this chapter's mapping of "camp." The icons of the legend occur within a column that becomes a middle zone between the institutional meaning of camp procedures on the left and the particularities of camping activities on the right. The National Park Service developed the former designations with the cooperation of the United Nations for an internationally legible iconographic code. The terminology on the left provides an archetypal set that remains generic in its universality, and the reinterpretation of the icon on the right associates specific practices with the camping iconography.[14] The National Park Service further institutionalized the icons in the late 1960s and early 1970s as "picture symbols" with standardized color schemes and with an allowance for variations only if they were accepted universally. As open-ended symbols, these icons function as indexes of a particular place and allow for multiple interpretations.

In this indexical mode of "pointing," the legendary icons are also tropes for spatial practice. The disconnect between the Park Service's institutionalization of camp spaces and the everyday experience of camping partially furnishes this figurative aspect. These tropes also create and mediate a dialogic relationship between the images and the text. From within this middle column, the tropes register both the figural treatment of camp and the often-nonstandard marginalized usage of space in practiced camping. The middle columnar zone also serves conceptually as a site for marginal annotations that might arise from the reader's parallel practices of placemaking. Such usage and diagrammatic possibility derive from the practice of embellishing the spaces between text of sixteenth-century manuscripts with iconographic words and phrases that were called "tropes." This figural marginalia resonates with the peripheral and annotative practice of camping in its habitual reinvention of spaces—whether textual, theoretical, or environmental. As such, the icons also operate as sites of association.[15]

Campfire

We begin around the campfire with its evocations of reverie. The flickering space of the campfire combines reflection, revelation, and regeneration. For Wallace Stevens, the idea of fire in Florida's context combines fleeting poetic imagery with introspective reflection. In the poem "Nomad Exquisite," Stevens concludes: "So, in me, come flinging / Forms, flames, and the flakes of flames."[16] Stevens composed the original version of this poem on the back of a postcard sent on January 15, 1919, from West Palm Beach, Florida, to Harriet Monroe, who had recently founded *Poetry* magazine.[17] In the earlier postcard draft, the last line reads "Fruits, forms, flowers, flakes, and fountains," and the image on the postcard includes a grove of palms in the front of a "dimly perceptible house."[18] In the later version, Stevens maintained the alliteration but has converted the memory of these associations into elements of fire. As an object, fire forms the central zone of the idealized campground. In camping, fire is an object of reverie that makes up the "center" of a starlike formation around which associations, memories, and fragmented ideas are gathered and assembled.[19] The camp and its surrounding context illuminated by the flicker of flames make up an oneiric landscape.[20] Memories of past campfires precede setting up camp. Campfires, legends, and stories begin an unfolding that is complemented by the siting of camp and its attributes of place-making. Camping stories extend, branch, and loop back with the campfire as the epicenter of reverie.[21]

The circumstantial properties of fire recur in the process of situating camp in its paradoxical context. The "centering" on the object of fire yields contemplation: "Fire is for the man who is contemplating it an example of a sudden change or development and an example of a circumstantial development."[22] The campfire becomes the site of storytelling—a collective of campers and their myths. And in a religious setting, fire intensifies the transcendent and sublime qualities of the camp. At camp meetings and revivals, the campfire's illumination combined with the flicker of small candle flames and a nineteenth-century religious "fascination with light" to create an "unreal atmosphere."[23] On a larger natural scale, fire in Florida

Campfire at Airstream rally. Courtesy of Airstream Corporation.

serves a regenerative purpose to "clear" the natural understory and at the same time to enrich the soil for new growth. As in Stevens's poem, the Florida landscape is fire, and campfires reference this natural occurrence at the human scale.

Tales of natural beauty, unmatched economic prospects, and buried treasure make up Florida's legend. Stories in the WPA guide reveal that, as early as the sixteenth century, Florida was known as "Stolida, the Land of Fools." The guide also includes a story told to a friend in an English tavern by an inebriated sailor who boasts of "perles in oysteres" and "glysterynge gold [sold] for tryfels."[24] Gibsonton's story of a hidden cache of pirate's gold illustrates the problematic reading of maps without legends when the orientation is lost by an overeager resident:

> GIBSONTON, 11.2 m. (250 pop.), a small trailer camp and filling station town on the southern bank of the Alafia River, was named for the pioneer Gibson family. Residents have often searched for buried pirate gold in the vicinity. One group, in possession of an old chart, unearthed a skeleton sitting upright, and below it a metal disk with

> the points of the compass and a needle marked on its face; in the excitement one of the party snatched up the compass without noting the direction indicated by the needle. Although many days were spent excavating the premises, no treasure was found.[25]

Campfire Stories

Following Highway 41 ten miles south of Tampa, we reach Gibsonton on the south bank of the Alafia River.[26] Named for the pioneer Gibson family, the town has become the winter home for many sideshow organizers and performers. Carnival trailers and trailer-homes radiate out from the centrally located Giant's Camp Restaurant at the Alafia Bridge, where in the early 1920s Eddie and Grace LaMay, who would become the town's first carnival performer residents, stopped their trailer to rest and fish. After this accidental beginning, "Gibtown" became the preferred home of the nation's carnival performer community. Gibsonton was home to Melvin Burkhart, the original "Human Blockhead," before his death in 2001. The town has the only post office in the country with a special counter for midgets and a museum of grave dirt collected from around the world.

Located along McKay Bay south of Tampa's Ybor City, De Soto Park provided the temporal and liminal territory for the first meeting of the Tin Can Tourists of the World along the park's "Easy Street" in 1919. Subsequent gatherings were held biannually in Michigan and in Tampa or Sarasota until the club's disbanding in 1977. The origin of the term "Tin Can" is the topic of a legendary dispute among "Tin Canners." Attributions for the name given to the group vary—an allusion to "Tin Lizzie," a derivation from the tinned food cans that provided sustenance on the road, and a reference to the gasoline and water cans welded to the sides of the travelers' cars and trailers for extra storage capacity.[27] In spite of these disparate stories, a tin can welded to an automobile's radiator symbolized membership and solidarity to all road-weary "Tin Canners."

Ruskin is twenty miles south of Tampa along Highway 41. The town was founded by George M. Miller as a socialist colony based on the philosophy of Englishman John Ruskin. A Chicago lawyer and educator, Miller had previously attempted similar communities in Missouri and Michigan. With this Florida incarnation, Miller decided to construct a self-sustaining agricultural and educational colony. The new town's activities revolved around

Ruskin College, with its curriculum modeled on that of Oxford University. A 1919 fire, which according to legend was started by a professor smoking a pipe, destroyed all of the campus's wooden buildings. Once known as America's "Salad Bowl," today Ruskin is considered the "Tomato Capital of the World," a distinction it celebrates with its Tomato Festival and Parade.[28]

Forming the northern edge of this region, Masaryktown is ten miles south of Brooksville and thirty-six miles north of Tampa and is named for Thomas G. Masaryk, resistance leader during World War I and the first president of Czechoslovakia. In 1924, at the height of the Florida real estate "Big Boom," Joseph Joscak, an editor of a Czechoslovakian newspaper serving New York City, formed the Hernando Plantation Company and purchased twenty-four thousand acres of land in Hernando and Pasco Counties, with the intention of developing an agricultural colony for Czechoslovakian immigrants.[29] By December of that year, 135 shareholders had reached, by train and then by bus, what was called "Joscak's Paradise":

> For several years Masaryktown was culturally isolated, for the settlers spoke only the mother tongue. Although the younger generation is quite Americanized, the farming methods of Europe are still employed, and Old World feast days are celebrated. On Czechoslovakian Independence Day, October 28, the man for whom the village was named is toasted and honored. Some phase of the republic's history is told in a pageant, and special services are held throughout the day at the church. Folk dances conclude the festivities. The women wear wide flaring skirts, with bustle effects, and handmade bonnets of intricate pattern. Enlivened by native wines, all Masaryktown dances to the strains of Bohemian music.[30]

Breaking Camp

The process begins again by breaking camp and continues with a rereading of the procedure itself. In legends, this reading is a gathering of fragments that are assembled to generate myth.[31] The narration of these component parts requires an elision of missing facts, unsubstantiated anecdotes, and unexplained phenomena. In contrast, breaking camp involves disassembly, an analytical mode of reading that seeks to document rather than mythologize. Unpacking meanings of "camp" begins with the linguistic "vessel" itself—iconographically linked to the image of the pick-up truck with its camper top that is folded and unfolded. Breaking camp is an activity of simultaneously dismantling into component parts and folding up into a unit. Supportive of camp's complexities, such analysis unfolds the relations of fragments and does not necessarily reduce to essences. To break camp is to utilize all possible interstitial space.[32]

The process of defining camp yields a cluster of meanings that include contradictory degrees of permanence and organization. Related to its designation as a field prepared for games or leveled as a function of battle, the abstraction of camp as an open space or field of settlement approximates the blank, empty pages of John Ruskin's own journals and diaries, which according to Jay Fellows are the "original models for the space of his world."[33] In Fellows's interpretation, Ruskin's pages are spaces to be filled. Closely linked is the indexical space of Ruskin's journals and museums: "The Index World is a world of appetite, clouds, and dense space, as convoluted as a labyrinth without discernible beginning or end . . . the intricate infinity of the Index World predicts the blank pages of silence and madness."[34] Related to "camp space," Ruskinian blank space does not result in an emptying out but instead operates as an interval or spatial moment between the episodic experience of texts and places. More directly linked to the built environment itself is the Ruskin College campus. Here, the campus space is the physical layout of the original turpentine camp and is also the experience of the campus's in-between spaces.

The photographs of Walker Evans and Ernest Meyer document these in-between zones of camp space. The photographers' documentary mode registers camp's multivalences. The images portray both entrenched and exodical space—that is, camps that have either gained a degree of permanence or are clearly temporary structures of a limited duration. In the Evans and Meyers photographs, camp becomes a "sublime ruin of abandonment" and a momentary pause at the interval of breaking camp. Having recently completed *Let Us Now Praise Famous Men* with James Agee, Evans was hired in 1941 by United Press International president Karl Bickel to take photographs to illustrate *The Mangrove Coast,* a book about the natural beauty and mythology of the Gulf Coast of Florida.[35]

The photographs of Ernest Meyer, though portraying camp space with

Walker Evans, Trailer in camp, Sarasota, Florida, 1941. Gelatin silver. 6 31/32" x 6 13/32". The J. Paul Getty Museum, Los Angeles.

less evidence of social criticism, document the events from the road and within the campground. His travels with the Tin Can Tourists yielded an array of images depicting everyday life in the camp—from proud displays of fish catches to suppertime gatherings, from broad family portraits to games of cards. Improvised arrays of fabric shelters drape across the assembly-line automobile forms to create ad hoc domestic places that transform spaces of mass production into Bedouin-like zones of temporary occupation. The vehicles' mobility is resisted and tied down by the tensile canvas, rope, and wire. These camp spaces are in a perpetual process of lingering or waiting as opposed to signaling a desire to depart. "Breaking" in this sense is a remaining until the last possible moment before an eventual untying. As "nomadic hesitation," this propensity to linger collapses, or at least compresses, the cyclic sequence of camping.[36] The operations of "breaking" and "siting" camp are woven together by the reciprocal activities of packing and unpacking.

Siting Camp

Siting camp initiates the writing of a camp practice. The implications of transportability and ephemerality, wrapped up in its fabric material and symbolized in the tepee-tent icon, frame problems inherent in camping's layered histories. As previously noted, the temporary quality of the campground occupation leaves limited traces for study. Established before the town of Ruskin and Ruskin College, the Shell Point Hotel was built on a Calusa Indian shell midden. After the hotel's destruction by a hurricane, the shells of the mound were spread to form Ruskin's roads, and today a single archived photograph testifies to the original site's existence. Such layering and redistribution characterize the siting of camp in its habitual practice. A historical stratification, often traceless but multilayered, defines the transience of the Florida landscape. The Spanish missions in the northern part of the state predate those of the western United States.[37] Although none of the Floridian mission structures or native constructions remains intact, enough information has been gleaned from archaeological investigations to reconstruct the Mission San Luis de Apalachee in present-day Tallahassee.[38] The region's early and contemporary histories are equally altered, covered, and mixed. Natural disasters or transformations through fire, wind, rain, and hurricane coexist with economic expedience, climatic variability, and material decay.

Siting and resiting camp in Florida thus involves both recollection and re-collection. Recollecting yields a siting of camp *next to* a monument (memorializing) or *over* a previous installation (layering memory). In this sense, camps are sited para-sitically *next to* and palimpsestically *over.* Re-collecting is siting as reassembly—a "gathering together" that includes the negotiation of the components of camping with the potential site's situation.[39] In each of these two cases, the siting of camp includes the paradox of proximity and distance, both spatially and theoretically. In the situation of camping, such contradiction can result in the establishment of a supplemental space between the ideal and the real.

Associated with the camping practice, the icons also portray a parasitical ground, which occurs between the real and the copy. Charles Sanders Peirce describes this circumstance as the moment when the difference between real and copy fades away and the diagram, whether it is an image or another work of art, becomes the "very thing" itself—an unexpectedly realized "icon."[40] The contemplation of an icon and the practice of camping share a momentarily dreamlike condition. Siting camp invites Peirce's sense of "pure dream" in that it precedes the extended time of thoughtful observation after camp has been made. This initial state or condition can be compared to Bachelard's "oneiric experience" of the landscape, in which the campfire becomes the focus of contemplation and the archetypal fire becomes the object of reverie. Following Peirce's logic, the "pure dream" of camping occurs between the particularity of place and the recollection, or understanding, of previously experienced places in general. In the expediency of siting a temporary occupation, particular features of a place are often elided. Contours are flattened and distances are condensed while more general features of past camps are retrieved. The commonality of tree heights and canopy, the orientations of sunrise and sunset, or the access to the campsite might be universalized. At the same time, absolutely unique landscape features and site conditions are often considered at the expense of generalized knowledge about previous sitings of camps.

Campsites, like those photographed by Evans and Meyer, are usually found at a periphery, tangential to a line of movement, or within gray-area urban and suburban zones. Varying degrees of monumentality often inscribe and identify these sites. Ironically, the Giant's Camp Restaurant shares its location at Gibsonton's social nexus with the gravestone business Rock's Monuments owned by Judy Tomaini, the daughter of the famed sideshow performers Al the "Giant" and Jeanie the "Half-Girl." In some cases, campsites are associated with the monumentality of a natural landscape feature, which in Florida might be a spring or sinkhole, giving a fixed point of interest.[41] However, zones and vectors hold a greater importance in discussing Florida's camp space: the siting of camp traverses these territories, moving vectorially from place to place. Only with hesitation, or pause, does siting become a provisional site. Travelers along Highway 41 stop for lunch and end up founding Gibsonton, and the Tin Can Tourists are evicted from Tampa and must locate a territory that they themselves can oversee. At the regional scale of the southeastern United States at the southern edge of the Deep South, central Florida is a frontier space that has remained "open" to various types of settlement. Openness and centrifugality also operate *within* central Florida's peninsular condition, dispersing campsites to liminal locations along bays, rivers, and roadsides. As a result of the habitual nature of camping, the settlers' occupations of the marginalized territories of Masaryktown, Ruskin, and Gibsonton were overlays onto other camps.

Masaryktown was founded around a logging camp. Its functioning sawmill provided the timber for the construction of the original hotel. Gibsonton has its origins in a fish camp along the Alafia River. George Miller and the other founders of the town of Ruskin took over an abandoned turpentine camp, with its three pine slab cabins serving as classrooms and dormitories for the college campus and its commissary store providing a schoolhouse for the area. And the Tin Can Tourists had their first meeting in Tampa's De Soto Park, which as a result soon became an active municipal campground.

The siting of these camps results from a drive to utopia. Dislocated camp communities rely on new methods for inventing spaces and inhabiting places: "liberation of an alienated culture is bound to take (or attempt to take) utopian forms—precisely because there are no known models."[42] The camping mentality includes this impulse or desire for something else—again locating camp between the ideal and the real—the Ruskinian zone of "nothing but process": "I work down or up to my mark, and let the reader see process and progress, not caring to conceal them, but this book [*Proserpina*] will be nothing but process."[43] If camping is a fugitive process, then camp space is a utopian model, partially realized in these aspirational places—a kind of environmental "jottings." Ruskin's writing procedure and the sitings of Florida's camps recall Susan Sontag's structuring of her text on camp in which notations record the phenomenon's "fugitive sensibility."[44] Not surprisingly, in her "Notes on Camp" of 1964, Sontag identifies John Ruskin and his work as falling within what she calls "camp."[45]

Florida's camp, with its utopic impulse, occupies a middle ground between heterotopia and a residual atopia. Camp's negotiation of paradoxes of identity and anonymity, place and placelessness, and stability and movement oscillates between atopia's nonplace and heterotopia's network of juxtaposed places.[46] Both intentional and seemingly unintentional places define atopias of camping.[47] Camp space, which includes qualities of atopia, heterotopia, and utopia, answers the rhetorical query, in its uneasy coordination of a nonplace and a place of belonging: "it would be interesting to discuss what kinds of compatibilities are possible between the principles of identity and belonging . . . and the principles of atopia."[48] Campers identify with a particular place through the siting of camp. And camp space, as a mobile family space or, more broadly, as a community space, accommodates another kind of identity, regardless of the specificity of place.

Subcultures such as the Tin Can Tourists must reconfigure an ideal "home" at a distance from their real home between each arrival and departure. At times, this planned impermanence translates into a more stable, locative settling down. In Masaryktown, a condition of the foreigner-at-home, a corollary of the Tin Canners' tourist-at-home, exists in which the utopic impulse is to create an encampment or enclave within the fissures of political and cultural influence, found in the Floridian frontier. With her term "tourist-at-home" collapsing the distinctions of the visitor and the visited, the touristic and the domestic, Lucy Lippard contends that the appropriation of modes of tourism allows a critical artistic practice at the interstice of the local and the foreign.[49] In Ruskin, Florida, George Miller hoped to create a self-sustaining community that synthesized John Ruskin's socialist ideals concerning education with manual labor and to avoid earlier attempts that had failed because of their geographic and economic proximity to existing communities—in Missouri with the town of Trenton, and in Michigan with Glen Ellyn, a suburb of Chicago.

Clearing Camp

Although occupying the fissures between institutional and civic structures in Florida's landscape, camping is delimited and does generate codes of practice and structure. The preliminary activity for making a place within the camping sequence is a grasping—an immediate gestural extension of space—which is then to be followed by the process of cultivating. The architect and theorist Ignasi de Solà-Morales revises Heidegger's discussion of *Raum*, in which space is made room for, and instead talks about space as making room for an event, what he calls the "architecture of the event."[50] Semipermanent forms of encampment come about from this cultivation stage, as opposed to the more temporal and transitory nature of the initial stage of siting and grasping. The clearing of camp occurs between these two activities. Camping is gestural in the way the camper expresses intentionality within space and is "agricultural" in the establishment of temporary lineaments. Performance precedes the organization of a cultivated location in the former. This condition of clearing also helps differentiate camping and campsite.

This in-between situation can best be described in the term "camp(site)." The practice of camping holds camp as place and camp as idea—the parenthetical (site) allows for nodes of location and sitedness within the overall process. Though varying in degrees of stability, both procedures of siting and grasping characterize camping and encompass what amounts to a dwelling or moving along a surface—a scraping rather than a foundational or hierarchical stratification. This characteristic of "surfacing" relates to the fact that the campsite's history is found in traces, whether ruins or the discarded remains of the campfire. In the legend above, the icon of the pedestrian "trail" highlights this idea of temporary traces and tracks. The question that remains is whether these fragmentary relics can provide a basis for the writing of a flexible code, symbolized in the dashed line of the trail. "Clearing" proposes a form of ichnography, literally "track-writing," that serves, like the legend's icons, as a reservoir of possibilities from which to generate a fluid taxonomy for camp spaces.[51]

In the absence of governmental or other regulatory enforcement, an array of much more subtle codes of ownership and separation exist in ad hoc trailer camps and early mobile home parks. Often these "codes" outline expected behavior and delimit permitted quotidian activities rather than inscribe overtly legalistic regulations, which are commonplace in many planned communities. An extreme example of the latter situation is Celebration, Florida, in which visible design and furnishing elements are mandated by the town's restrictive codes.[52] Similarly, the New Urbanist community of Seaside utilizes two codes, urban and architectural, to regulate the appearance of its development. This intentioned homogeneity contrasts with the campground's codification.

Camp codes restrict certain forms of behavior, from garbage management to driving speed to loud music, but are more permissive in terms of a heterogeneous physical and built environment. The visual field is not aestheticized and for the most part remains a personal response within a social milieu. The rules of camping provide a "code of conduct" that affects actions but does not directly dictate the setting or situation. The informality of this code is evident in a handwritten text for Gulf Hills Campground outlining what are identified as "a few rules."[53] The campground's owners, Ray and Wilma Warren, provided this text to me in 1992 upon my agreeing to rent a mobile home in northwest Florida along the Gulf of Mexico. It is ironic that the informal rules are written on a yellow legal pad. The implications that this document has for the architecture of the camp include the unscripted appearance of the grounds, the "open" nature of the space of the camp as a park ("This is an adult Park . . ."), and the ad hoc hybridity of materials and modes of construction.

As in the rules for Gulf Hills Campground, a new grammar of physical boundaries applies to these spaces, which are not divided into the series of thresholds found in more permanent developments like typical suburban areas. Instead, the camp spaces exist as one large threshold, in which there is a hierarchical flattening out, a general equivalence of objects—folding lawn chairs, plastic tables, automobiles, and work-related tools.[54] Unencumbered by radical shifts in climate or topography, Florida camps, whether RV, trailer, or mobile home parks, remain open without extensive partitioning or enclosure. In the case of Gibsonton's varied landscape, chain-link fences serve as porous and transparent containers for these collections. But other edges and boundaries are not merely visually transgressed—curbs are driven over, and grass blends to sand, and gravel becomes pavement. And rules can be handwritten on yellow legal pads, on what appears to be a case-by-case basis, as with the Gulf Hills Campground "code."

One source for codification, apart from the quotidian stipulations of the Gulf Hills code, is the definition of the moral character of prospective members of the society. The implicit principle of social respect within trailer camps was concretized and verbalized in the settlements of Ruskin and by the members of the Tin Can Tourists. The town of Ruskin's socialism was built on membership to the Commongood Society. Based on the 3 *H*s—the enrichment of Head, Heart, and Hands—the society held regular meetings and oversaw all facets of community life from land-parceling to prescribing college curriculum. The Tin Can Tourists similarly defined "moral character" as a necessary qualification of applicants. Their "Ode" coming out of the 1930s also reflects a set of shared principles, at times sounding the battle hymn and resonating with a lively camaraderie.[55]

In Gibsonton, the zoning laws themselves were manipulated to allow for the carnival performers to store the tools of their trade in their front yards. The special classification "Residential Show Business," an overlay on the Hillsborough County Land Use Plan, allows for concession stands, performing animals, Ferris wheels, and any other apparatus associated with the carnival or circus to be exhibited in the resident's yard. In revised versions of the Hillsborough County Land Use Plan, the zoning classification has been termed "SB" or "Show Business Overlay."[56] Sarasota, previously the focus of circus and carnival life, changed its zoning laws in the 1950s to disallow such exhibitions.[57] Whether zoned or not, this inversion of front and back is typical of camp space.

Making Camp

Making camp involves a negotiation of disparate elements. These oppositions include distance and proximity, mobility and fixity, inconsequential and grand, standard and nonstandard, local and global, and permanent and impermanent. As shown in constructing this idea of camp, it follows that Florida vernacular is a state or condition of flux spatially and temporally—a labile language, the grammar of which operates between these oppositional qualities. This grammar occupies the space between the words in the phrase "mobile home."[58]

Camps borrow from what is near and far, local and foreign. Ruskin College's campus included a later building supplemental to its wood-plank structures. Built in 1912, the President's Home served not only as George Miller's dwelling but also as additional classrooms. The permanence of its construction contrasts with the more temporary qualities of the original turpentine camp buildings that did not survive the fire of 1919. The inspiration for this building's design came directly from a sketch by John Ruskin in his *Poetry of Architecture* published in 1893. For the college, Adaline Dickman Miller adapted what Ruskin had termed a "Swiss Cottage." Thus, added to the Ruskin camp, this construction negotiates a paradoxical permanence, collaged stylistically from its remote and absent "founder's" sketches.

The semipermanent buildings of a campground often serve as meeting halls, restrooms, and changing rooms. In Masaryktown, the hotel building served as the original community lodge. And ironically, today the community center is housed in a prefabricated building dressed in a thin layer of facing stone. Serving a similar purpose as the focal point of the carnival community in Gibsonton, the Giant's Camp Restaurant is also sheathed in this material to simulate a masonry or stone wall. As with the stone facing on these public buildings, domestic mobility, also fleeting in its temporality, is disguised by transforming mobile homes and trailers. Skirting around the base of mobile homes hides the vehicles' wheels and mimics foundation walls. Materials for defining territories or properties elicit this paradox; chain-link fences become the primary permanent delineation of space within the semipermanent campsite.

Making camp has aspects of collecting before an acknowledged dispersal. It is this centripetal-centrifugal dilemma that characterizes the Federal Writers' Project's documentation of these towns. The *Guide to the Southernmost State* was at one and the same time meant as a guidebook for tourists and other travelers and as a reference book for local Floridians—a cento of fragmentary bits of culture, folklore, architecture, and history allowing for an uncanny evaluation of disparate elements in a common work. As a result, seemingly unrelated elements occur in proximity to each other: song lyrics next to descriptions of agricultural research work next to accounts of fishing spots next to depictions of local architecture and construction materials. Similarly, John Ruskin's Guild of St. George, which served as one of the models of Ruskin, Florida, combined an external gathering of disparate objects placed in a traveling museum exhibit. This museum of objects was paralleled by a process of internal documentation and reformulation in Ruskin's written projects—a florid, even lavish, style of writing and collecting, using dashes to prolong a breathless stream of movement that at any moment could be diverted by a swerve of digression. In his writing and imagination, Ruskin moved easily between Venice and Manchester and among painting, sculpture, architecture, and philosophy. It is this patchwork mentality that links Ruskin to camp in all of its indeterminacy and paradox.

Camp becomes a methodology. This use of camping as a method for research and a possible architectural practice relates to Michel de Certeau's work on walking as a critical act. De Certeau proposed a metaphorical city that exists as a network of fissures and interstitial spaces between totalizing institutional constructions.[59] The operation of walking, like camping, is based on the idea that these spaces, within de Certeau's "network of antidiscipline," can be understood as "practiced place."[60] Camp is thus a mode for practicing paradox and inventing spaces for researching and building within Florida's context.

Each camp is an event registered in its siting, its clearing, its oral communication of codes, and its continual making, down to the storied ashes of its reused fire pit. Making camp remains a way of negotiating space that allows for territories of smoothness between regulated spaces. This does not say that camps lack regulation, but the "perimeter projects" of camp space subvert and often invert typical ways of space-making in a much more subtle operational mode.[61] John Ruskin's parenthetical statements in his writing point toward themes of digression and accumulation, which summarize the broken cycle of the camping process. In one case from the manuscript of the *Praeterita*, Ruskin noted that he would return to a particular location in the narrative at a later time. Editors of his collected *Works* later countered Ruskin's paradoxically vague specificity ("I will return . . .") with the subsequent annotation to the parenthetical note—"This, however, was not done."[62] This pairing of statements yields a diagram with an open-ended meaning that lies along the horizontal dividing line between parenthetical statement and footnote:

(I will return to this point afterwards[1])

---

[1][This, however, was not done.][63]

The fugitive nature of camp disallows the promise of a locative return "to this point." In the making of camp (site) remains parenthetical in the possibility of a return to the same place.[64] Although Ruskin does not return to this precise location in the text, he does revisit its ideas throughout his work. It is this purposeful meandering, a fragile structure, that characterizes both Ruskin's project and that of camping. Ruskin's work is a milieu in which process generates the form as opposed to the complete formalization of process. Ruskin's parenthetical statements occur in the "peripheral middle" and the "delaying middle," where the making of camp also happens in all of its material, spatial, and metaphorical significance.[65] Ruskin's patchwork of references characterizes a "mosaic" of digression and discursiveness within this "peripheral middle." And the Floridian campsite may be understood not only as a phenomenon of betweenness but also as a model for looking into the territorialization and reterritorialization of contemporary space and place.[66] Closely linked to camp's paradoxes and operative modes, nomadism exists as a residual "intermezzo" in which points of stability exist only as vectorial starts and finishes, leads and trajectories—amounting to a series of "local absolutes."[67] Contextualized within the Deep South, camp space is characterized by this sense of "nonlimited locality" allowing for a region defined as much by paradox as geography.[68]

Florida camping in particular and southern camping in general rely on paradox and marginality for their capacities of place-making. Ruskin, Masaryktown, and Gibsonton exist as enduringly speculative productions of place unique to Florida but are at the same time symptomatic of the intensive camping procedures found elsewhere. In this way, camp resonates with de Solà-Morales's understanding of place as "not a ground, keeping faith with certain images, . . . [but] rather a conjectural foundation, a ritual of and in time, capable of fixing a point of particular intensity."[69] The Tin Can Tourists will begin this intensive cycle of places with their first organized camping experience on a cool December day in 1919, within the municipal ground of De Soto Park.

Chapter 6

MOVING IMAGES OF HOME

Tin Can Tourism and Florida's Municipal Camps

Early Xmas morn we all went out to Mr. Macklin's (about 12 ½ mile in the country) . . . got us a big dinner of wild duck and beef roast. . . . But that's not all, we had all the oranges we could eat and also grapefruit and tangerines. . . . [W]e all had our pictures taken after dinner. . . . Right after supper we talked a while then went home (I mean back to "Desota Park").

—RUTH DEERING (December 25, 1921)

Siting the Autocamp: Para-siting the Municipal

On Christmas Day of 1921, Ruth Deering confuses her home with a municipal tourist camp south of Tampa's Ybor City. Her experiences in the camp and at Mr. Macklin's occur at the height of the Tin Can Tourists' tenure in Florida's burgeoning coastal city. Having first set up camp in December 1919, the Tourists will soon be evicted from their municipal home-away-from-home as a result of their changing relationship to the citizens of Ybor City. Meanwhile, the exotic provision of oranges, grapefruit, and tangerines during the winter's holiday highlights other unlikely convergences of variability with fixity and proximity with distance. In this fragmentary account, we understand the narrator's identification of the public camp with home—a site from which her family will then travel the measured distance out to the country.[1] The municipal camp itself has become the outsider's surrogate home, a base for holiday celebrations and family time. This camp is a place momentarily embraced by the city but at the same time held at a distance, with the understanding that it is an ephemeral home to be remembered through "pictures taken after dinner."

In its most contemporary meaning, "municipal" denotes administration by a local governing town or city. Such management is characteristically particular to the specific political entity, and its rules are outlined as legal policy or municipal law. More distinctively, the municipal classification covers internal affairs as opposed to external, or international, relations. Similarly, the municipal can designate that which belongs to one place only.[2] In Florida's municipal camps, the introduction of camping tourists tends to transform this extreme localization. Ruth Deering's excitement over the bountiful citrus at Christmas supper identifies her family as nonlocals who bring diverse backgrounds, external to the municipality. The concept of the municipal is also related to the Roman *municipium*, which identified town or city with some inhabitants of Roman citizenship but also with laws that were internally defined rather than being determined by imperial mandate. From a distance, Rome still exacted duties (*muni*) from their captured or otherwise "taken" (*capere*) outlying towns and cities (*municipia*).

The Roman *municipium* was distinct from the ancient Empire's *colonia* even though both types of cities fall within the general understanding of "colony." This important differentiation reflects the complexities of the Tin Can Tourists' relationship to early twentieth-century cities in Florida and the rest of the nation. With a higher degree of connection to Rome and thus more rights of citizenship accorded its inhabitants, the Roman *colonia* settlements were considered higher-level outposts of the Empire than the *municipia*. Technically, *colonia* were composed of resettled Roman citizens, primarily veteran soldiers and retired legionnaires who were granted land in these foreign, newly conquered, and hostile areas. The Romans had initially derived the term *colonia* from the Greek word ἀποικοι—meaning "people from home."

As the municipal camps grew in popularity, particularly among the Tin Can Tourists, the *municipia-colonia* distinction came to light in terms of class differences and the acceptable lengths of camping tenure. In the host city's view, the middle-class camping tourist of the 1920s occupied a *colonia* while the hobo tourist came to occupy the camp as a *municipia* that, without adequate control, could be lost in a chronic vagrancy. In the eyes of the citizen, the former group came to spend money and the latter quickly became an administrative burden.

The third level in the hierarchy of Roman cities was identified as *civitas*, which were initially used by the Romans as relatively independent settlements to administer Gaulic or Celtic tribes in the distant northwest region of the Empire. At first, *civitas* was applied to the Celtic tribes and their traditional territories of influence, but the term later identified the principal towns of these areas. *Civitas* offers a model for understanding the Tourist camps themselves in relation to other forms of administration (municipality and colony) and in terms of self-regulating and organizing tendencies of the campers themselves. Forming a mobile submunicipality, the camping tourists organized themselves as a *civitas,* based not on a particular location but on a larger, more open-ended region. This flexible *civitas,* associated with the Tin Can Tourists of the World, can be set up within a municipality, be dismantled, and then be reformed in another location. Each permutation of *civitas* borrows from the locality's municipal structure but at the same time remains independent with its own set of rules and its characteristics of mobility. Para-siting a municipality, each camping settlement becomes a self-regulated municipal colony.[3]

The municipal campgrounds, while not offering residents voting rights, mirror the municipal-colonial relation between Rome and *municipia* at a smaller scale and in a closer physical proximity. Whether embedded within the relatively urban commercial zone or in an outlying annexed area more evocative of the pastoral landscape, the camps themselves reinterpret the *municipia* from which the concept of the municipal originated. With their external status amplified by public perception of the citizens, the residents of the municipal campground in effect become foreigners-at-home, or "people away from home."[4]

Municipal camps had been created with the proliferation of autocampers as a viable means of tourism. Previously, tourists camped in the yards of farmers who could offer space for pitching tents and water for drinking and washing. Charles Treffert, one of the original Tin Can Tourists who appeared in a 1948 *Life* magazine, noted that in the early years before more established campgrounds were readily available, "you camped anywhere you could find water—farm, schoolhouse, graveyard."[5] Such informal campgrounds relied on road access; and as the number of autocampers grew, roadsides and farm lots could no longer support the influx of tourists. In their early formulations, the municipality provided alternative campgrounds for free, with the expectation that tourists would spend money in the towns or cities where the camps were sited. These municipal camps av-

Ernest Meyer, Early roadside autocamp, south of Jacksonville, Florida, 1922. Courtesy of the Library of Florida History and the Florida Historical Society, Cocoa, Fla.

eraged in size from ten to fifteen acres and often included kitchens, lavatories, and showers as well as provisions for laundry, lights and other electrical services. Funded by both government chambers of commerce and local businesses, advertisements from the period show how towns competed for the tourism dollars expected from these camps, which were the source of civic pride and rivalry between towns and cities.

Following the rapid increase in production and availability of the automobile, municipal camps grew rapidly from around 1915 to 1925. By 1918, in response to tourists' trepidation in traveling to the Deep South, Henry McNair, editor of the American Automobile Association's guidebooks, declared that "a sortie through the southland was no longer considered a perilous adventure."[6] Between 1920 and 1924, three to six thousand free municipal autocamps could be found throughout the United States.[7] During this period, Kenneth Lewis Roberts, correspondent for the *Saturday Evening Post,* made his initial trip to Florida and wrote his first articles. Surveying the early trailer life of the Tin Can Tourists with articles titled "The Time-Killers" and "The Sun-Hunters," Roberts publicized an exaggerated

Tin Can Tourists in De Soto Park, 1920. State Library and Archives of Florida.

profile of Florida's economic potential and helped initiate the Florida boom of the 1920s with his far-fetched claims.[8] At this relatively early stage in autocamping culture, the West and the North housed the greatest distribution of these camps. In 1921, Florida's share of the camping culture was limited to thirty-eight established autocamps, of which thirty fell under the authority of municipalities.[9] By the late 1930s, as roads across the southern United States were improved, the distribution of trailer parks had increased in Florida, giving the state, along with California, the greatest concentration of campgrounds in the country. Produced in 1938 by *Automobile and Trailer Magazine* to show the distribution of trailer camps, a map of the United States clearly represents the highest density of camps surrounding the Tampa Bay area and Miami.[10]

In spite of the overall growth of the nation's camping culture, the decline of the municipal campground began in 1924 as the "vacationing camper" was differentiated from the "hobo tourist."[11] Postcards from the time period contrast the Tin Can Tourists' understanding of the freedom offered by a life on the road with the Florida public's scrutiny of the Tourists as a wild, unkempt group of frontiersmen. Illustrating the first set mainly derived from the tourists' perspective, one postcard shows Tin Can Tourist families camped along a road with autocampers, tents, and livestock forming an ad hoc settlement within the semitropical landscape of palm trees and Spanish moss. The caption "Where the Iceman cuts no ice—or Gas, or the Coal, or the Rent-man, either" references the camper's freedom from economic ties to a conventional societal structure. Other postcards present the Tin Can Tourist as a solitary figure with shotgun and tent and without motorized transportation. Cloaked in a hunter's outfit and partially obscured by the uncultivated natural landscape, this type of camper represented the stranger and the outsider.

Community resistance to the Tin Can Tourists was based on the idea that the free camps attracted a lower class of itinerants who did not contribute to the local economy and who stayed longer than their middle-class counterparts because of unemployment rather than leisure time. The terminology contrasting "camper" and "tourist" and "vacationer" and "hobo" indicates perceptions of the time. It was assumed in this period that the camper would remain for a short duration while the tourist's stay did not have a definite end. As far as purpose of the visit, the vacationer would return to a particular place of origin, and the hobo did not have a home and thus made the camp home.[12]

As a result of this growing resentment, city-run camps began charging fees, and private camps replaced the free municipal camps. In addition, most camps set limits on the length of stay allowed, with the typical range falling between one and two weeks. The increase in the number of motels at this time also contributed to the decline of the free municipal camp. Early versions of the motel were the low-cost cabin and the cottage. Mary Anne Beecher places the advent of these more static campground constructions around 1925 as tourists sought "more comfort, convenience, and privacy."[13] The rise in popularity of the roadside motel-cottage weathered the Depression as a low-cost alternative to hotels and as easily accessible lodging for short trips. In the early 1930s, an estimated four hundred thousand cabins had been built since 1929.[14] The small scale and relative simplicity in the construction of the cabin and the cottage made this building type competitive with the campgrounds that primarily provided open space for autocampers and automobiles with tents.

This availability of a semipermanent option resulted in a hierarchy of lodgings for the tourist population arriving in Florida. In 1924, the City of West Palm Beach created a campground to supplement the camping spaces in one of its city parks that had become overcrowded. Known as Bacon Park, the new camp was regulated by city ordinance; and in 1926, four thousand campers generated more than $15,000 of revenue each month.[15] In the park, the City of West Palm Beach provided three choices for camping, each primarily defined by its sheltered space and degree of permanence: "a bungalow, a tent house, or a wooden-floor tent. The most expensive of these, a comfortable fourteen-by-twenty-eight-foot bungalow rented for a reasonable sixty-five dollars a month. In addition to these bungalows were sixty-eight 'close to nature' tent houses . . . smaller and designed to suit motorists on tighter budgets. For even less money—five dollars a week—visitors . . . could rent one of a hundred tents [set] on wooden floors."[16]

These class and income differentiations heightened negative perceptions of lower- and middle-income groups such as the Tin Can Tourists. By the mid-1930s, tents were no longer allowed in many campsites, and camps were designated as "parks" in the late 1930s to distinguish the grounds from those patronized by "hobo tourists."[17] *Trailer Tintypes,* a cartoon series published in the *New York Herald Tribune* during this time, reflects the internal class structure of trailer life in general and of Tin Can Tourist culture in particular. Authored by an artist known as "Webster," the cartoons portray conversations among campers, conflicts with trailer park owners, and feigned nostalgia for a distant home in an atmosphere of tedious leisure. In one exemplary sketch titled *The Aristocrat,* a tourist peeling potatoes rhetorically mulls over the difficult choice of her next luxurious vacation stop, while seated outside of her trailer. This cartoon was included as a newspaper clipping in a Tin Can Tourist scrapbook compiled by the Levett family and housed at the Florida State Archives.[18]

In spite of the aristocratic aspirations of some *Tintype* characters, Tin Can Tourists were generally "sun-hunters" as opposed to the more affluent and less pecuniary "time-killers" of the upper-class resorts of Miami, Palm Beach, and Daytona. However, the sun-hunter demographic does include a mix of professions and incomes: "The sun-hunters are not recruited from any one class of citizen . . . there are some bankers among them—and some burglars. . . . The bulk of them are farmers. Next to them come contractors, builders, and carpenters. The sun-hunters are the people who can get away from home with the least amount of trouble . . . among them one finds retired businessmen of all sorts, dairymen, doctors, bankers, lawyers."[19] While the general public's view of the Tin Can Tourists in the 1920s and 1930s may have been that the group was comprised of a lower class of vacationers, the Tin Can Tourists were a professionally diverse group, predominantly middle-class vacationers with backgrounds reflecting the nation's vast prewar working population.[20]

Between Denizen and Citizen: Siting the Tin Can Tourists of the World

The Tin Can Tourists first met in December 1919 on the grounds of De Soto Park on McKay Bay in Tampa. Twenty-one autocampers led by James M. Morrison of Chicago convened at the park and in January 1920 produced a constitution and bylaws.[21] In the winter of 1921–22, De Soto Park hosted 4,329 camping tourists, who arrived in 1,571 autocampers.[22] The density of the park's accommodations is evident in Ernest Meyer's photographs from that time. Meyer went on to document the life of the Tin Can Tourists on the road, traveling throughout Florida with his wife, Jennie, and beloved cat between 1921 and 1924.[23]

From its inception, the Tin Can Tourists appended the clause "of the World" to their organization's name. The irony of this designation can be found in their relatively insular habitation of the central Florida region, with occasional spring and summer conventions in Michigan and the Midwest. Although the Tin Can Tourists' aspirations of global camping were not realized, their invocation of the "world" as their range does point to their uniquely conditional tenancy within their own country. The Tourists are neither denizens nor citizens; they are foreigners-at-home. In its original meaning, denizen denoted both the original inhabitant of a place and a foreigner admitted to provisional residency within a particular district. The denizen thus has the dual, if contradictory, status as indweller and resident foreigner.[24] The Tin Can Tourists spent nearly half of the calendar year away from their original home. The typical "season" for the campers was from November to early April, and was punctuated by "homecomings" and conventions—originally Dade City in November, Arcadia for Christmas, and Sarasota or Tampa from January to April.[25] Summer conventions

Panoramic view of Payne Park and Tin Can Tourist convention at Payne Park and City Trailer Camp, Sarasota, Florida, 1936. State Library and Archives of Florida.

were held in Michigan, and, according to the Federal Writers' Project, the Tin Can Tourists came to "live all or much of the year in trailers or house cars."[26] As a recently revitalized organization, the Tin Can Tourists continue their camping traditions across the country and through the virtual geography of their Web site, realizing their original aspirations of globality through an active Internet community.

The Tin Can Tourists' organization adopted a position for its members that was a combination of the denizen's indwelling and impermanence and the citizen's native and relatively stable standing. In this chosen status, the Tourists effectively became foreigners who did not dwell outside the limits of their indigenous home country but well within it. Here, the foreigner-at-home, as Lucy Lippard's "tourist-at-home," navigates the vaguely familiar and the slightly out-of-place phenomena of particular places. From the perspectives of both visitor and visited, tourism then becomes a way of experiencing the "disparate surfaces of everyday life" and eventually "reintegrating our fragmented world."[27] This blurred ground of objectivity and subjectivity offered by the tourist-at-home, though contrasting with the involuntary nature of immigrant or exiled status, does complicate the Tin Can Tourists' status as tourists. The Tourists, in their drive to self-determination, have chosen a condition of exile within the political and economic structure of their country. For them, the exiled state is a self-imposed absence from their home and aspects of their home country. This absence can be understood as a dislocation generated from within.

While not renouncing their citizenship or the policies of their homeland outright, these quasi-expatriates become immigrants who arrive within the previously known but temporarily forgotten places of their original home.[28]

In the case of the Tin Can Tourists, this internal displacement is a chosen circumstance that yields a domestic "worldliness" in which its members, as a part of a mythologized seasonal tour, are habitually returning home—whether this is the original home or a newly established home. Although for the most part not by his own choice and out of his control, Ulysses' own peregrinations culminated in his return home disguised as a beggar, only to reveal himself to the unsuspecting suitors as the rightful, though temporarily displaced, home owner. Without Athena's mandate that the "exile must return," this foreigner-at-home condition is carried on through the Tin Can Tourists in their seasonally activated transformations of home. The activities of the Tin Can Tourists thus occur between chronic arrival and departure—a constant moving between homes, a residual homecoming. Accordingly, the "tourists," while neither exiles nor immigrants, are truly in a state of "becoming."

Not until Wally Byam's projects of the 1950s and 1960s was camping attempted on a global scale. As founder of the Airstream company and his Caravanners Club International (WBCCI), Byam led caravans around the world not only to engage media attention for his company but also to carry out his doctrine of unfettered leisure and discovery—principles that read like a utopic manifesto of mobility. Byam's Creed includes the following objectives: "To place the great wide world at your doorstep for you who yearn to travel with all the comforts of home. . . . To open a whole world of new experiences. . . . To encourage clubs . . . that provide an endless source of . . . personal expression. . . . To lead caravans wherever the four winds blow . . . over twinkling boulevards, across trackless deserts . . . to the traveled and untraveled corners of the earth. . . . To play some part in promoting international goodwill . . . through person-to-person contact."[29] Byam's trips included the Central American Caravan of 1956 from the United States to Managua, Nicaragua; the African Caravan of 1959 from Cape Town to Cairo; and the Caravan around the World of 1963 from Lisbon to Tokyo.[30] His vision included a network of stopping points, or rest areas, across the world to facilitate Airstream owners in their discovery of "new experiences."[31] These "Land Yacht Harbors" were conceived as minimal mobile home parks with facilities for refueling, temporarily occupying a location, and witnessing a particular place.[32] Although more socialistic and less proprietary, remarks delivered by the president of the World Community of

Game of checkers at Airstream rally. Courtesy of Airstream Corporation.

Gypsies in 1963 reflect Byam's aspirations for global camping: "We are the living symbols of a world without frontiers, a world of freedom, without weapons, where each man may travel without let or hindrance from the steppes of central Asia to the Atlantic coast, from the high plateaus of South Africa to the forests of Finland."[33]

Byam's creed and his overall vision also resonates with Constant Nieuwenhuys's outline for a culture of New Babylon. Constant's design for an experimental Gypsy camp inspired his subsequent development of the utopian city New Babylon. Also derived from his studies of Gypsy culture and from the 1963 address by the president of the World Community of Gypsies in particular were Constant's writings about the practice (as opposed to the simple idea) of freedom: "It is obvious that a person free to use his time for the whole of his life, free to go where he wants, when he wants, cannot make the greatest use of his freedom in a world ruled by the clock and the imperative of a fixed abode."[34] Practicing freedom requires the mobilization of home. But the ideal dweller within New Babylon does not just have the freedom to move from place to place—he or she has the need and the right to play. Constant and other Situationist readings of Johan Huizinga's ludic formulation resonate with the Byam creed: "As a way of life Homo Ludens will demand, firstly, that he responds to his need for playing, for adventure, for mobility, as well as the conditions that facilitate the free creation of his own life."[35] Practitioners of recreational camping, from the Tin Can Tourists to Wally Byam's Airstreamers, espouse this reciprocal necessity for creative adventure: to play is to be free, and freeing sites for play allows us to create, to live.

Originally published in *Trailer Travel* magazine, Wally Byam's "Four Freedoms" echo many of the ideas expressed in Constant's work. This liberating creed includes freedom from arrangements, freedom from the problems of age, freedom to know, and freedom for fun. Freedom from arrangements is made possible by having "all your accommodations right there with you" such that home is where you come to rest and stop. According to Byam, freedom to know can occur because of the combination of mobility and home—when "you travel in a trailer, you meet people in their homes and they meet you in theirs." Finally, Byam concludes with the fourth freedom, which is the culmination of the other three—freedom for fun allows the trailerite "to relax and lose yourself mentally."[36] Byam's emphasis on freedom is ultimately framed within the possibility for ludic activity.[37]

While the Tin Can Tourists' organization itself allows for a placelessness today within the global medium of the Internet,[38] the pragmatics of access and leisure require siting the global campground adjacent to or within an infrastructural system. A series of central Florida locations reflect components of this practical connection between camp and city. In the 1920s, De Soto Park was linked to Ybor City and Tampa's commercial core by a trolley line. The park's proximity to McKay Bay also provided water access for recreational activities such as fishing and boating. Another camp, the Municipal Trailer Park on Oregon Avenue, was centrally located on the Hillsborough River within Tampa's early city limits. This park was also the site for the Convention Hall, built between 1946 and 1947 by the Work Projects Administration to provide space for convention proceedings, dances, and exhibitions. Farther to the south, the Sarasota City Trailer Camp was located in the larger sports and entertainment grounds called Payne Park.[39] Immediately adjacent to this permanent camp, which has been identified as the first of its kind in Florida, was a baseball diamond for minor league baseball and spring practice for major league players.[40] Contiguous with the town's early fairgrounds site, this area had been originally laid out and zoned as "public properties" in John Nolen's 1924 "Comprehensive City Plan" for Sarasota.[41] In the open area between the gridded trailer park and the baseball stadium, the Tin Can Tourists camped and held many of their annual conventions, including the 13th Annual Convention in 1936.

A diverse set of material gives us glimpses into the experiences and place-making activities of the Tin Can Tourists of the World as they camped out in Florida. Postcards, aerial and documentary photography, and early camping manuals contemporary with the Tourists' early years are themselves "sites" for reading and rewriting the camping practices of the first part of the twentieth century.[42] The postcards of the Tin Can Tourists from the 1920s and 1930s were derived and created from early photographs that documented everyday life of the camp. In addition to this documentation, the postcards, as devices of communication, contain information about their use by tourists and about relationships to external or distant places in their written content and addresses. The postcard thus reflects a mobile construct through which multiple places are read, much like the autocamping vehicles themselves. Aerial photography is used to reconstruct the plans of the campsites—serving as one means of siting the Tin Can Tourist camps. Such surveys use the aerial photograph, the historical bird's-eye view, and the cadastral map to image, read, and understand early camps. Such images include historic aerial photographs from the 1930s and 1940s and recurring views of Payne Park from Ringling Towers in Sarasota.

In a series that includes a photograph from the Ringling Towers, Marion Post Wolcott provided an important documentary record of the Sarasota City Trailer Camp adjacent to Payne Park. Her photographic documentation describes the camping environment of the Tin Can Tourists and contrasts significantly with her photographs of the migrant camps near Lake Okeechobee.[43] Transcending the documentary format, Wolcott's photographs become an "act of making" that tells the story of a particular day in a specific campsite. As a resource in the difficult study of built phenomena that are as ephemeral and relatively undocumented as camps—no longer existing or constantly transformed, the "photograph as act of making" is an important joint between the historiographic reading of the recorded image and its use to draw conclusions and carry out a mapping of the campsites.

Marion Post Wolcott, "Guests at Sarasota trailer park, Sarasota, Florida," January 1941. Nitrate negative. Library of Congress, Prints and Photographs Division, FSA-OWI Collection.

In such documentation, to photograph is not to "take the world as object" but to construct and actively make it: "Every photographic image is a story, at once insufficient yet real, a take, a discharging on something that does not allow itself to be used up once and for all because it will always retain its fleeting, ungraspable condition."[44]

Serving as research evidence and forming methodological threads, these types of documentation can be understood in three ways: as historical objects of study, as acts of photographic place-making, and as methods for mapping the space of the camps. Combined with these documentary methods, Elon Jessup's *Motor Camping Book* provides a series of operations for camping and making a home on the road. Published in 1921, Jessup's manual recorded the techniques used by autocampers, including the Tin Can Tourists, to convert mass-produced vehicles to improvised autocampers, erecting tent structures attached to automobiles, and generally to make camp within the early municipal and roadside grounds. These techniques and procedures outline both historical and contemporary camping practices and prepared itinerants of that time to "clear camp."

Clearing Camp: Denizen as Citizen

> Communities have a real responsibility in the provision of camping facilities. . . . There is so much to be said about a so-called nomad population in the nation, using trailers as the "modus-operandi." . . . Trailerites form the great middle class of our people. . . . Consequently, communities should provide for them with the finest facilities possible.
>
> —J. E. KAVANAUGH (December 1936)

Debate over the regulation of camps, both municipal and private, grew as campgrounds became more accessible and thus more frequented by tourists. The field for debate included ad hoc planning, self-regulation, and, ultimately, legal contests. Early camps were for the most part unregulated, spontaneous, and dispersed. The openness of pasturelands and the informality of roadsides and schoolhouse or churchyard lots allowed an abundance of choices for positioning the campsite. Within early autocampgrounds, this informality of location was continued, relying equally on alternative features such as shade trees, water accessibility, and campfire pits. The relation among individual campsites was resolved on the ground, and each new siting of camp responded to the preexisting conditions. This arrangement is evident in the photographs of Ernest Meyer and other early Tin Can Tourist photographers. As tents and autocampers became campers and trailers pulled by cars and as camping vehicles increased in size, a gridded order was imposed on the camp, and trailer lots became inscribed on the camping ground. By the early 1930s, campgrounds such as the Sarasota City Trailer Camp used a grid plan to organize large areas of ground and territory.

As the physical layout of these municipally run camps evolved, ordering principles of particular camping organizations were recast to establish suitable camping practice and legitimacy in the public eye. Ratified in 1920, the constitution of the Tin Can Tourists was an early example of self-regulation by a camping group. The organization's constitution and bylaws included sections about assisting other campers, fastening the Tin Can Tourist emblem (a tin can) to the autocamper, and leaving clean grounds. Invoking overtly religious language, the latter set of rules included the broad admonition: "to spread the gospel of cleanliness to all camps, as well as enforce the rules governing all public campgrounds."[45] In the interest of self-preservation and in response to community resistance to trailer camps, the trailer industry picked up on this idea of autonomous governance in the mid-1930s. The industry's leaders believed that communities had a responsibility to provide for the "trailerites [as] the great middle-class of our people." The industry also argued that some form of standardization of camping practices must be introduced to sustain the camping population's respectability. Such industry involvement in the regulation of trailers and their use resulted in the gradual replacement of the ad hoc home-modified autocampers with mass-produced and commercially manufactured camping units.

Questions of permanence have implications for how camps are understood in terms of dwelling and the "clearing" of settlement. In spite of the efforts of camping organizations and the trailer industry, communities resisted campgrounds on the basis of the increasing permanence of the camping vehicles and the trailer owners' reluctance to pay taxes. Here, the point of contention was no longer the itinerancy of the Tourists but their inveteracy and determination to remain in one place for extended periods of time. Early legal decisions attempted to explain the new type of accommodations found in trailers and campers by defining "dwelling." In *People v. Gumarsol*, the City of Orchard Lake, Michigan, sued Hillard Gumarsol for violating an ordinance that required dwellings to be at least four hundred square feet. Gumarsol, along with five other trailer owners, had parked his travel trailer on a rented lot in the summer of 1936. Rather than moving his trailer at the end of the season, Gumarsol placed it on blocks, attached a porch, and prepared it for his return the next summer. The court ruled in favor of the City. In the court documents, Justice Green wrote: "It is the opinion of this court that a house trailer of the type occupied by the defendant and having a great many appointments of a modern home would come under the scope of a human dwelling whether it stands upon blocks or the wheels attached thereto or whether it be coupled to or detached from an automobile."[46] According to the court, the ad hoc foundation of concrete blocks did not affect how the trailer's use was defined. Because Gumarsol exceeded the typical limits of seasonal tenure by an itinerant resident, the trailer was a dwelling. Its permanency was a function of its extended duration. And its living area, less than two hundred square feet, made it an inadequate and illegal dwelling.

Such a ruling, which was typical for the time, resulted in a problematic relation between the legal definitions of the "modern home" and the requirements of local codes. Because of this disconnect, the trailers were legally permanent and thus subject to taxation; but in their permanence, they violated building code stipulations that in the previous case required a minimum area of living space.[47] Subsequent legal debates have attempted to differentiate vehicle and dwelling, to regulate the potential permanence of trailers by limiting the time of residence, and to define what qualifies as an attachment. In most rulings, the mobile unit of the trailer and its owner-inhabitants have been considered "transient" while the land on which camping occurs has a permanent use classification. This ambiguity does not hinder many municipalities from taxing the trailers either as vehicles, by assessing registration fees, or as more permanent dwellings, by levying property taxes. To avoid taxation and land use restrictions, trailer parks and campgrounds quickly moved outside municipal boundaries, thus maintaining access to the municipality's attractions and infrastructure while evading regulation. As cities expanded, camps that fell within the annexed area were usually included in the incorporation and were allowed to maintain their status through "existing use" grandfathering.[48]

With the host-guest model used as a framework for studying relationships within the campsite itself, the trailer site is the zone of the camper as guest within the host territory of the campground. Here, the varying degrees of attachment occur primarily in a sectional connection with the ground. For the Tin Can Tourists, attachment works vertically from the ground to the establishment of shelter through operations of unfolding, binding, wrapping, stretching, adding, and unpacking.[49] In Gibsonton, the idea of lightness "grounding" the camp will be understood as stacking, laying, and resting in the support of temporary dwellings and the midway structures of the Florida State Fair.[50] In contrast to the sectional role of maintaining and transforming that occurs within the campsite, the *positioning* of the trailers, RVs, and mobile homes is predominantly a question of arrangement and fit—relations typically understood in plan. The ambiguity of legal judgments about what constitutes a "temporary residence" fails to clarify the long-held assumption that "proper dwellings are *attached* to land, whether or not that land is owned by [the] occupant." The difficulties in defining, adjudicating, and regulating the actual degree of attachment to the ground result in far-reaching zoning regulations that force camps to the urban periphery, only later to be reincorporated into the expanding municipality.[51] Historically and legalistically, attachment denotes an apprehending, detaining, and controlling. This derivation is closer to the idea of "attack" than the early meaning and components of "attach."[52] The root of attachment is in the word "tack"—a nail that is used to fix something temporarily, with the assumption that it will be removed. The problem of attachment is this ever-present conditional and future detachment.

Making Camp: The Bricoleur's (Mobile) Home Laboratory

> To architects, the house has the appeal of the experiment. In smaller, more compact and controlled situations it becomes possible to speculate. The house becomes a laboratory for ideas.
>
> —BEATRIZ COLOMINA (1998)

Taking this idea of the house as a site of architectural speculation, we might say that the trailer has also been a medium of experimentation for both architects and home owners. In such an exercise, the scale and mobility of the trailer is simultaneously liberating and restrictive. The autocamper's small scale allows an individual either to modify and add onto an existing vehicle or to construct a trailer that can be towed. Each case requires an inventive use of available space and surface. In this case, attachment includes joining materials of wood, canvas, and steel as well as connecting to the ground. The mobility of the trailer necessitates that the attachments are permanently fixed to resist road wear and wind shear, can be detached and stored, or are self-contained within the surface or volume of a towed unit.

Early designs of trailers created by inventors and entrepreneurs were often either appropriated for mass production or developed as prototypes for commercial sale by the inventor. Glenn Curtiss's goose-neck designs of 1920 and experiments with the tent trailer in 1921 evolved into the Curtiss Aerocar of 1929.[53] William Hawley Bowlus's early monocoque designs of the 1920s were also eventually transformed into Wally Byam's Airstream, begun in 1936. Arthur Sherman's early garage trials resulted in the creation of the Covered Wagon brand of trailer design in 1929. At a larger scale, William Stout's folding house of 1938 later became the post–World War II double-wide mobile home. In addition, do-it-yourself plans for building trailers could be ordered from companies and were published in magazines such as *Popular Mechanics* and books like A. Frederick Collins's *How to Build a Motor Car Trailer*, with its first edition in 1936. An early version of the Airstream trailer designed by Wally Byam appeared in a 1935 issue of *Popular Mechanics*.

In their engagement of the making process, these trailers became mobile laboratories of home. Edward Casey has outlined this homesite of making: "The *Compars* of the immured state, its closed-in and regularized spatiality, is starkly etched against the *Dispars* of the bricoleur's home laboratory—which, like a transitory nomadic camp, is set up with materials ready at hand in a casually arranged workplace that lacks fortified walls."[54] Early autocampers and trailers and the camps that they made relied on a knowledge of material forces rather than a strict interest in material forms and quantifiable economics or efficiencies. For example, in their expandability and changeability, the tent shelters of the Tin Can Tourists responded both to specific sitings and to the configuration of the parked automobile itself. In Gibsonton, we will see that the domestic yard is a cleared site for the arrangement of cars, trailers, and carnival equipment, while the municipal camp provided a public territory for the grounding of leisure-time experimentation.[55] It is important to note however that in Casey's conceptual construction, these laboratories maintain a degree of permanence while the early boxcars and autocampers remained mobile. The "situatedness" created by a lack of economic means is not present in these middle-class constructs that are inherently movable because of the owner's expendable income for repairs and fuel as well as the time for leisure and tourism.

Although the autocampers, trailers, and more contemporary mobile homes share origins in industrialized production, the way the autocampers are modified differs from the changes made to trailers that have stopped moving. This difference between the two modes of construction affects the built environment of the camp. Where semifixed trailers and mobile homes grow and are modified for the most part by an externalized accretion, the transformation of autocampers occurs from the inside out with minor exte-

Home-devised sheet metal, running-board food box, which unfolds to serve as a cooking area. The camper is filling his gasoline stove. Elon Jessup, *The Motor Camping Book,* G. P. Putnam's Sons, 1921.

rior additions. Without a stable ground for support, autocamper structure relies on the bearing capacity of the axles and the main chassis. Elements must be cantilevered from this central axial support. Additional shelter, then, must either fold out from or be stored within the camper. Canvas and fabric made up the material for the latter configuration; and in many early camps, the tentlike shelters were built off of the automobile's shell. The tent was fused with the motorcar. This synthesis of fabric and steel in the early autocamper made a house and homesite through the material domestication of a mass-produced vehicle.

Making Home: An Abbreviated Operational Manual for Tin Can Tourism

Camping manuals of the 1920s and 1930s often provided detailed accounts of camping shelter types and cooking kits and made recommendations for modifying camping vehicles. In *The Motor Camping Book,* Elon Jessup presented techniques of combining the tent and the autocamper as shelters for the camper's home on the road. In the 1920s, the tent was the most economical and readily available camping equipment for travelers such as the Tin Can Tourists. The tents would soon be formalized in more complex fold-out campers and would eventually be superseded by the travel trailers of the 1930s. In spite of its less frequent use in modern campgrounds, the tent has remained an important part of understanding camping practices. And Henry David Thoreau's discussions attribute many of the tent's features to the understanding of home in early American housing. Thoreau describes the crystallization of his original tent around him in his Walden dwelling:

> The only house I had been the owner of before, if I except a boat, was a tent, which I used occasionally when making excursions in the summer, and this is still rolled up in my garret. . . . With this more substantial shelter about me, I had made some progress toward settling in the world. This frame, so slightly clad, was a sort of crystallization around me, and reacted on the builder. It was suggestive somewhat as a picture in outlines. I did not need to go out doors to take the air, for the atmosphere within had lost none of its freshness. It was not so much within doors as behind a door where I sat, even in the rainiest weather.[56]

Crystallizing home to house is an excursive movement that combines with the inhabitant's sense of atmosphere and temporary shelter. The sheltering process "reacts on" the dweller. Thoreau delineates the reactive framework and his "settling" in the world as a progression of operations. The crystallization moves outward from the tent's inhabitant and inward from his natural surroundings, from which he is "behind the door."

Jessup's camping manual also outlines a crystallization process, relating the tent and the autocamping house. The following operations form a preliminary manual for "making home" on the road:

Unfolding

Jessup uses the example of the Auto-Kamp trailer to describe how a tent assembly can unfold into a bedroom, dining room, and kitchen: "When ready to make camp, the trailer itself becomes your home . . . practically a portable house, ready for light housekeeping."[57] The camping trailer folds out to provide the framework for two double beds that are separated by a floor space, which is the "bed" of the towed trailer, with a central table that folds up. The sides of the bed are used as benches to sit at the table. Two central poles flip up to form the ridge of the canvas tent (see illustrations on pp. 228 and 229).

Attaching/Binding

Attaching is used in the lean-to and the one-legged tent that can be attached to the side of the car. In this version of the combination bed and tent, the running board supports the head rail, and the car top serves as the ridge pole. As the single, centralized leg to support the bed, a stretcher rail attached to the car's foot rail is notched to allow height adjustments. Two outriggers staked to the ground angle away from the tent's foot end and keep the entire assemblage in tension. Jessup notes that this tent-bed avoids the problem of other configurations that tend to be "drawn out of center" by eccentric forces. Other collapsible tents that combine bed with tent and utilize folded storage on the running board are the Stoll, the Schilling, and the Andersen. The running-board canvas bed can serve as a table when the tent is removed. Other configurations of car-supported tents, used for shelter as opposed to bedding, include the single-tent lean-to, the single-tent with car shelter (what Jessup calls the "portable garage"), and the double tent in which the car is completely enclosed or "wrapped" in canvas.[58]

Opening

The procedure of attaching can allow for opening. The architect R. M. Schindler follows Thoreau's progression from tent to house in his design for the Bennati Cabin. This A-frame house orients its open end to a domesticated fireplace and a more rustic fire-hearth beyond. On the same axis, the other end of the frame opens to the garage. Schindler's interpretation of the autocamping layout—fire, shelter, vehicle—also opens to frame the natural surroundings (see illustrations on p. 230).[59]

Wrapping

Fabric wraps the automobile for protection and comfort. With the slip-on cover as in the Burch double outfit, one canvas covers the hood and the other protects the rear of the automobile. In the double tent, the automobile is completely obscured, and the occupant must enter at the middle where the automobile offers the most support for the canvas wrapping (see illustrations on pp. 87 and 117).

Stretching

The running-board canvas bed extends into the landscape through a process of stretching. Here, stretching is not only a "stretching out" but also a preparing to extend the limbs of the body, in a reclining or prostrate arrangement.[60] With the weight of the automobile as anchor, the running-board bed is a prestretched hammock that enfolds the camper (see illustration on p. 228). See also the faceted wedge-style tent in "Attaching."

Adding

Additive operations utilize preexisting tent styles independent of the motor car. These types of tenting include the wall tent, the marquee tent with its front flap attached to the side of the car top, the unfaceted wedge tent in which the car top serves as a support for one end of the wedge tent, and the miner tent. This latter tent is a type regionally specific to the western plains. Reminiscent of the tepee, its pyramidal form has less headroom than the marquee tent but is very good at resisting wind deflection as a result of its form and an often externalized tripod framework.[61]

Boxing: Storing and Unpacking the Box

With the running-board box, Elon Jessup documents in detail two "well-constructed" boxes that he has encountered in his own camping excursions. Presented as a part of the camping guidebook, his specifications in each case provide the necessary information for readers to complete their own versions of the invention. Outlining components of one design, Jessup describes a homemade sheet-metal running-board food box, which was "stationed on the left running board; the depth was that of the running board, it was as high as the top of the tonneau door, and the width extended from the rear end of the running board to the front seat door."[62] The sheets of galvanized iron are riveted together, and the front sheet is detachable and can be used as a table with two steel rods to support its outside end. After relating the camper's use of felt lining to keep dirt and dust out of the food storage, Jessup describes in detail the interior of the portable cupboard, which "contained two shelves holding foodstuffs, mostly in jars. Between the jars were small wood partitions which . . . were loose and detachable so that if one wished to rearrange the jars, this could easily be done."[63]

In another camping outfit outlined in Jessup's text, the right running board carries beds, folding chairs, and other small pieces of equipment; and the rear of the automobile stores a lean-to tent. The author again focuses on the left running-board compartment: "It is 50 inches long, thus extending practically the full length of the running board and 26 inches high. The bottom is 12 inches wide while the top is only 9 inches wide. In general appearance, the outfit when closed reminds one of an upright piano box."[64] This operation of *boxing* is also found in the camping boxcar. Writing on economy, Henry David Thoreau proposes a railroad box as economical shelter: "Consider first how slight a shelter is absolutely necessary. . . . I used to see a large box by the railroad, six feet long by three wide, in which the laborers locked up their tools at night, and it suggested to me that every man who was hard pushed might get such a one for a dollar. . . . You could sit up as late as you pleased, and, whenever you got up, go abroad without any landlord or houselord dogging you for rent."[65]

Like the trailer in its portability and the portable cupboard in its economy of space, the railroad box serves as a heuristic in Thoreau's discourse on a personal freedom unhampered by landlords, rent, and debt. The Tin Can Tourists worked from a similar kind of freedom.[66] The postcard with a photograph of the tourists from the 1920s, stating prominently that the camp is a place where "the iceman cuts no ice," shows that the Tin Can Tourist is not bound to the daily delivery of services as those encumbered by permanent dwelling. Thoreau echoes this burden of obtaining ice in his description of a unique camping practice that allows one to avoid such an extravagant need: "Whoever camps for a week in summer by the shore of a pond, needs only bury a pail of water a few feet deep in the shade of his camp to be independent of the luxury of ice." Continuing his argument for an economy of dwelling, Thoreau laments, "we now no longer camp as for a night, but have settled down on earth and forgotten heaven."[67] For Thoreau, the permanence of dwelling has sacrificed the human proximity to nature that exists in camps. Camping operations are not ends in themselves but are ways to reconnect with oneself and with nature, outside of the "dogged" vexations of the luxurious house. Unfolding, attaching, opening, stretching, and even boxing, the Tin Can Tourists were free to move from place to place and to connect with the "forgotten heaven."

Walking Camp / Making the Spaces in Between

In January 1939, Marion Post Wolcott documented the Sarasota City Trailer Camp adjacent to Payne Park. She photographed the spaces outside of and between trailers—the threshold space external to the trailer's internal volume. Within the sequence of her photographs, Wolcott has also documented how the camper moves from threshold to threshold—a series of spaces that are seamless rather than episodic. Begun atop the Ringling Tower from which she captured a bird's-eye view of the camp, Wolcott's itinerary worked up and down the corridors of the camp's parked vehicles and trailers. Wolcott's photographs also illustrate relationships between the body, the vessel of the mobilized homes, and the camp as a whole.[68] Passing the camper with her shell garden, Wolcott finds shuffleboard players and shaded card games before being greeted by the Lulofs, whose point of origin is Grand Haven, Michigan. The photographer's episodic sequence slows to converse with a relative who mends a jacket and with Mr. and Mrs. Lulof, enjoying the afternoon's winter sun. Wolcott walks by other groups clustered under awnings for games and meals before pausing in front of a leisurely conversation among seven men seated on folding camp chairs, their camp circle broken to open up on the trailer park's semipublic corridor. Also between automobile and trailer, she finds a camper unfolding his running-board extension, while another repairs a table. Trailer screen doors are swung wide, awning windows have been raised, jalousies opened, and improvised stoops are precisely located thresholds next to more hastily laid mats. Not quite Thoreau's completely externalized interior at Walden, but a series of quotidian activities and domestic lives reconfigured between autocamping vehicles. In a final sequence, a camper coaxes her pet dog to drink water and return to its chicken-wire pen—all within the contracted space between fabric, tent supports, and automobile. Carried out through the story told by Wolcott's captions and the sequence of photographs, this narrative follows the camera as camping "eye"—a wandering interlocutor as participatory as it is objective.[69] Walking, camping, and photographing collide in a practice to mediate the "betweenness" of the camp's place and time.

Breaking Camp

Wolcott's "city within a city" was self-sufficient as a social and recreational enclave, but its formal and infrastructural connection to Sarasota remained a necessity for its functioning. Campers and citizens could break this tenuous attachment at any moment. In its siting, the mobile construct of the autocamp can be understood as a place with differing degrees of attachment to the municipal entity. In its earliest iteration, the autocamp was associated with the infrastructure of the farm, with its accessibility to well water and availability of open space for setting up camp. From this agrarian setting, the autocamp's next type of attachment occurred within the municipal fabric as a result of the town's recognition of a potential economy derived from the provision of campsites. In later permutations, such as Braden Castle Park, discussed in the next chapter, attachments became tangential to and eventually detached from municipal infrastructure. Necessary connections to a regionalized infrastructure were maintained for such services as electrical power and transportation networks, but for the most part the camp became a city within a city, an enclave, or a satellite city completely outside municipal districts.

Read within this host-guest-parasite framework, the sets of relations form a series of systems. In each case, the parasite attaches to the relation established between host and guest. In the first system, the municipality parasited the proximity of city and camp. The campers-vacationers were seen as temporary inhabitants provided with all of the amenities that the municipality has to offer its own citizens. The campers are also given the key to the city. As Wally Byam boasts in *Trailer Travel Here and Abroad*, "I've got so many keys to so many cities, I'm going to start a museum just for keys." With the rejection of this model by citizens of the city, and in essence the municipality itself, the quasi-municipality of the camp takes the parasitic position and operates on the relationship between the city structure as host and its citizens as guests. In this way, the foreigner-at-home's camping practice parasites the service infrastructure rather than the physical infrastructure. In this series of events, operating on principles of fixedness, the city loses control over the space of the camp. In terms of the "time" of camping, the departure is not given and is thus understood as a duration rather than a temporality. Yet paradoxically, the city expunges the camps for their inhabitants' inclination to permanence and disallows the noncontrolled or nonregulated fixity of the camp.[70] The space of the camp is one of fluctuation and is primarily defined through the dynamics of the place and how it is "practiced."[71] In spite of the legal delineation of the campsite itself as public park, municipal campground, or private property, the camp's realization occurs only through the activity of camping. It is thus the absolute local that outlines the place rather than the relative global of the legal definition of space.

The time of camping, that of arrivals and departures, also characterizes the changing camp-municipality link. The time of the regulated fixity of the free municipal camp is the temporality of the two-week stay. In the case of Tampa's municipality, the space of De Soto Park represents a less regulated version of the municipal camp. Becoming codified in places such as Braden Castle Park, the time of the self-regulated "fixity" is the duration of intercalary arrivals and departures. This duration applies both to the seasonal occupations of the cottage owners in the park as well as the mobile home owners and trailer site renters of the park's other areas. It is duration that will allow breaking camp to be closely linked to siting camp for the Tin Can Tourists who relocate to Braden Castle.

De-camping: Municipal Evictions

The low-flung herd is moving on its way
In Tin Can Fords, high with brats

And mattresses and stoves that seem to sway
In sympathy with misplaced beds and slats.

Why they move? God only knows, but thanks
That Tampa has a respite from the doom

That year from her friendly life yanks
The Gulf Port City from the Tin Can tomb.

They come, each with a sardine in his hand;
They fling their tents on homestead, field and lawn;

They never buy but try to own the land—
Lord knows, we Tampans joy when they are gone.[72]

The eviction of the camping public from municipal grounds marked the end of the short-lived symbiotic relationship between outsiders and the localized political entity. While the facilities of the modestly sized public parks were strained by the seasonal influx of campers, it was the disappointment of local businesses and citizens with the frugality of the camping public's middle class that precipitated the closing of many camps. Although lodged within town limits, the modes of camping employed by groups such as the Tin Can Tourists allowed for an economic maximization of interstitial public space that condensed their physical impact on the urban environment. The tent structures and small scale of the autocamping units themselves proved remarkably adaptable to a variety of situations. However, questions of duration of stay, effects on property values, and enduring thrift of the tent-dwelling autocampers led to their expulsion beyond the "city walls." It was ultimately public opinion rather than infrastructural strain, except for isolated issues of health and sanitation, that defeated the openness of the municipal campground.

The politics of space led to the exclusion of the Tin Can Tourists from De Soto Park. Taking up the subject of homelessness, Rosalind Krauss links the homeless person and public space as "dual products of the spatioeconomic conflicts that constitute contemporary urban restructuring."[73] The homeless person does not introduce conflict into the space but is instead inextricably tied along with a proliferating array of other factors, such as the "public" of public space and perceptions of home. While the camping tourists were for the most part homeless by choice and thus could not be classified as such, the notion of a democratic spatial politics recognizes the fruitful rather than inimical significance of openness and difference.

> Fleet horses bear me
> Without fear or dismay
> Through distant places.
> And whoever sees me, knows me,
> And whoever knows me, calls me:
> The homeless man . . .
>
> No one dares to ask
> Me where my home is:
> Perhaps I have never
> Been fettered
> To space and the flying hours,
> Am as free as an eagle![74]

An alternative path that early city officials and citizens might have taken was to embrace rather than to expel the condition of difference introduced by the campers. It is my contention that understanding place provides another option for framing and understanding such conflictual situations. As an idea, place maintains an open field of exploration that is lost with the "closure" of spaces. Place, before an emplacement through this spatial closure, resonates with the early municipal camp in its allowance for the outsider to experience the open landscape of Florida. In their incompleteness, Florida's early towns were durable places of tourism that could be transformed with seasonal change. Florida's urban situation at this early stage suggests the possibility of an unromantic, nonnostalgic "placeful" utopia. Tied up in this paradox is E. B. White's vision of Florida and its rudimentary and "unfinished" cities of "bright cabanas, . . . dead sidewalks, and live jungle."[75] Though postdating the Tin Can Tourists' first incursion within Tampa's municipality, White's image of the ruins of speculation characterizes Florida's urban condition of the time. While such an "unfinished city" is in many ways the appropriate situation for urban camping, finding a way to allow for urban transience and openness, without incorporation or co-option, remains an important question for architects and planners today. At Braden Castle Park, we find an early version of the urbanization of camping, a condition that resulted from the impossibility of urban camping in Tampa, Florida.

Chapter 7

BRADEN CASTLE PARK

Eutopic Communities of Tourism

Siting Camp: March 1, 1924

The Civic Club of East Tampa forced the expulsion of the Tin Can Tourists from the municipality through legal action, closing De Soto Park to public camping on the first day of March 1924. An unpublished document titled "Origin and History of the Camping Tourists of America" in the collection of the Braden Castle Association shows that the idea to purchase land for a permanent camp was begun during the winter of 1921, when nineteen tourists camping at De Soto Park signed an agreement to purchase land, which would be platted and sold to other tourists. Three of the original signatories[1] to this agreement would later become residents of Braden Castle Park and would play an important role in its development. With increasing pressure from Tampa's citizenry, land purchase was again considered during the winter of 1922 without any result. In the summer of 1923, the Civic Club filed the initial lawsuit to close De Soto Park; however, the district's judge ordered that the park be opened on November 1 at its normal time. The Civic Club followed with another lawsuit filed in early 1924. This action proved successful, and the park was closed on March 1. In the introduction to his history of Braden Castle, H. E. Robbins cites conflict with residents of Ybor City as the impetus for Tampa mayor Perry G. Wall's expulsion of the tourists from De Soto Park; however, it appears that a general appeal from the community prompted the change in municipal regulations.[2]

Foreseeing the possibility of this closure, the members of the Tin Can

Tourists met on February 19, 1924, in the main pavilion of De Soto Park to discuss courses of action to address the imminent eviction. During the meeting, a motion submitted by Dr. V. M. George of Columbus, Ohio, led to the formation of a committee chaired by R. W. Vaughn of Rome City, Indiana, with the purpose of investigating the purchase of land for a tourist-owned campground. A subsequent meeting to choose committee members was held on February 20 in W. B. Jacobs's tent in De Soto Park.[3] The committee eventually incorporated as the Camping Tourists of America, with a board of directors that included Vaughn, W. J. Houck, Fred F. Bates, L. K. Supernaw, W. B. Jacobs, H. F. Wagner, and H. E. Robbins.[4] In this name change, the Camping Tourists sought both to distance themselves from the negative public perception of the Tin Can Tourists and to gain independence to look for a permanent campground site and thus avoid similar occurrences at other municipal camps. The bylaws of the Tin Can Tourists expressly disallowed property ownership by members in the camps occupied and patronized during the winter season.[5]

The closing of De Soto Park occurred one month earlier than was typical for the season, but the preparation for the eviction and the rapid purchase of land displaced the newly formed Camping Tourists of America for only one week. On March 8, 1924, the board members of the Camping Tourists of America agreed to purchase thirty-four acres of land known as the "Braden Castle Property" for $16,000, with funds received from 160 shareholders.[6] Located east of the town of Manatee, the property is defined by the confluence of the Braden and Manatee rivers; its southern section of land was originally interrupted by a tidal marsh that was later partitioned to form a landscaped lagoon.

Para-siting the Ruins / Parasiting History

The primary reason for the choice of the property in Manatee County was the existence of Braden Castle. Other factors included its immediate accessibility to water for transportation and recreation and nearby Bradenton's reputation for welcoming various forms of tourism. Known traditionally as the "Friendly City," Bradenton's promotional literature marketed the town's affability, and its Tourist Club included 1,567 registered members from thirty-eight states and three foreign countries in the 1929–30 season.[7]

Although already in ruins, Braden Castle's massive tabby walls connected the tourist group with a part of Florida history extending back into the mid-nineteenth century. The board of directors realized the importance of this link and quickly appointed H. E. Robbins to write a history of the site.[8] Robbins's account includes a prefatory poetic ode to the castle, a brief introduction, and an informal history of the grounds. Beginning with the line "Old Braden Castle is the tourist's home," the poem with its forced rhyme encapsulates Braden Castle Park's offerings from the "best in Florida's clime" to the area's "tonsorial artists" and the site's easy access to health care. The introduction reads as promotional literature, similarly touting freedom from domination by political influence and advertising up-to-date tourist camps with good fishing and facilities for dancing.[9]

The main text tells the history of the site from the perspective of the castle itself. In the narration, the castle witnesses the historical events of its siting, construction, and decay. Moreover, the castle's first-person chronicle testifies not only to Florida's early turbulent history but also to the Camping Tourists of America's regained legitimacy within Florida's landscape. This historical justification within Florida's cultural heritage is supplemented by the material solidity of the twenty-inch-thick tabby walls and their semblance and imagery of permanence, although slowly dissolving and crumbling. In the introduction, Robbins hyperbolically notes the uniqueness of the site: "[a] search in any part of the world will not disclose another Braden Castle, or any tourist-owned winter homes like these owned by two hundred share-holders and showing a good growth each year." In this account, Robbins later adds through the voice of the castle, "I shall see one of the greatest tourist homes in the world," and describes the Camping Tourists' occupation of the grounds as serving to both preserve and extend the castle's history by way of social and economic advancement. In the "History of Braden Castle," the castle continues the narration: "they wanted to keep me as a landmark, to show the tourists that would come here what this spot had experienced in the early history and the making of this part of the Land of Manatee." Robbins's purpose, along with other members of the board, is utopic in terms of this imagined place of limitless freedom and history, and he begins to reconstruct a mythology of the place by fabricating phrases such as "Land of Manatee."

Recording and rewriting the history of the castle, Dr. Robbins also cre-

Postcard of early tent camp at Braden Castle Park. Courtesy of Manatee County Clerk of Court Historical Records Library.

ates a foundational myth that is not unlike the stories of the area's original settlement in the nineteenth century. His family from Loudon County, Virginia, Dr. Joseph Addison Braden moved from Leon County in north Florida to the Manatee area in 1843 after the collapse of the Union Bank in Tallahassee and other deleterious economic fallout from the Panic of 1837. Having sold his heavily mortgaged tobacco plantation in north Florida, Braden began to establish a sugar plantation across the Manatee River from another Leon County transplant, Robert Gamble. By the late 1840s, Braden had secured approximately eleven hundred acres and had begun construction of the tabby walls for his plantation home.[10] Completed in 1851, the "castle" included two four-room levels with central hallways.[11] The house's footprint formed a fifty-foot square, with ten-foot-wide halls, twenty-foot-square rooms, and four chimneys serving eight fireplaces.

After the beginning of the Third Seminole War on December 17, 1855, the castle became a refuge for members of the surrounding community seeking security from the initial Seminole attack on March 31, 1856.[12] Ironically, the castle again served the purpose of security and refuge for the Camping Tourists of America, who retreated from municipal regulation in Tampa. As a result, the camp served as a strategic site of defense and resistance. The confluence of the Manatee and Braden rivers forms a point of land that partially protected the Braden residence's tenants from Indian attack. The park's central tidal pool, which was later modified to become "the lagoon," also contributed to the site's inaccessibility from land. With its marginal location at a distance from the central districts of Bradenton and the settlement of Manatee, the site afforded refuge and territorial distance for the tourists seeking a zone that could be self-regulated. The Braden house became identified as "Braden Castle" because of both its massive tabby construction and its use as a safe haven during frequent Indian attack.

The siting of Braden Castle Park at the nexus of the sugar plantation also engages, if unwittingly, the Camping Tourists in the memory of an economy with far-reaching territorial implications of power and land control. By collapsing the sugar plantation and the campground, the typical progression from camp to city, or from temporary to permanent, is reversed. The openness of the ruined plantation house provides a suitable site for the early camping activities, which are initially unenclosed, dispersed, and tied directly to the characteristics and topographies of the site's previous occupations. Before the arrival of the Camping Tourists, Manatee's residents frequented the castle grounds for weekend excursions, romantic walks, and family picnics.[13] Such romantic and idyllic scenography attracted the Camping Tourists, whose own historically temporal occupation of places resonated with the leisure activity and freedom of picnicking.[14] As the initial tents and trailers were replaced by the planning and construction of more permanent housing, the Tourists continued this tradition by setting aside the area around the ruins as a park. Early plans designate the communal space of this "park within a park" as the "Plaza." The siting of the camp in this location also works within the previous social-communal uses of the place; the plantation house itself had been not only a safe haven but also a focus of entertainment for the area's settlers. With its reduction in size to a property of 34 acres, Braden Castle Park is a contraction of the original 640-acre territory of the sugar plantation's section land claim. Today the ruins are no longer readily accessible for social gatherings and are fortified not by expansive land tenure but by a chain-link fence serving to protect the remnants of the crumbling tabby structures.

The siting of the park around the plantation house ruins elicits connections to the residual and the sublime. In an ironic coincidence of old and new, the community reformulated and thus "completed" the ruined domestic structure. As a historical remnant, the ruin gained a new meaning as the public plaza of Braden Castle Park. The ruined house was reinhabited along its periphery by way of the campers' circumscribed dwellings. Off-limits to actual habitation, the place of the ruin was reserved initially for special events by a locally generated communal agreement and later as a part of the district's oversight by state and federal regulations for historic preservation.[15] Consequently, the ruin was also utilized for its scenic effect, and thinking of the plantation site as a fragment allowed the campers effectively to inhabit the ruin. Because this habitation occurred at a remove from this central element, the ruins have become a scenographic element. And if the ruin is considered by the campers as a historical fragment in the passage of time, then psychologically the ruins invoke the powerful notion of time itself and the awe-inspiring power of the sublime. Braden Castle Park achieves a degree of timelessness in the archaeological collapsing of time through the combination of historical place and new building.

Clearing Camp: Braden Castle Park

> Those first on the ground proceeded to clear it off. A spot was cleared off north and west of the Castle where the camp was made. . . . Everyone took part in the work of clearing so that we soon had a large space cleared. . . . J. W. Dean, W. A. Harding, and Clem Neal were hired to clear up the land . . . in readiness for the allotment in the Fall. (March 10–11, 1924)
>
> —R. W. VAUGHN, "Origin of the Camping Tourists of America" (January 8, 1927)

Before siting the new campground at Braden Castle, the committee of Camping Tourists had placed an advertisement in the February 25, 1924 *Tampa Tribune* newspaper to see what campsites might be available for purchase.[16] With no responses from their advertisement and as a result of the immediacy of their eviction, the officers then visited potential sites between March 4 and 6.[17] Ironically, the real estate agents took the group of Camping Tourists to the comfortable environment of "Home Restaurant" for lunch during their evaluation of properties on March 4. Enthralled by the Braden Castle site, the group presented its findings to the board of directors on March 8: "the most available one [campsite] was the Braden Castle property, at Bradentown [*sic*], Florida, of thirty five acres, which was offered at sixteen thousand dollars ($16,000). A motion was made, seconded and carried that the site be purchased at the offer made. The said site being on the Manatee River in the city of Manatee, Florida."[18] On March 10, the officers reported that the purchase had been finalized, and the group decided to place another advertisement in the *Tampa Tribune* to attract investors and residents to the future camping community.[19] In addition, the group sought approval of a charter for its nonprofit corporation "to perfect an organization for the betterment of the camping tourists."[20] The application for charter was rejected by Tampa's governing officials, but was eventually accepted by Manatee County's administrators.[21]

Subsequent meetings of the Camping Tourists provided the basis for the writing of the bylaws, which then influenced the camp's organization and construction. On April 1, 1924, a motion was passed that "all political and religious meetings be referred to the director and incorporated into the by-laws."[22] A draft of the bylaws was read and discussed on April 3, and the officers moved to add the stipulation that "no established merchandising be allowed on Braden Castle campground."[23] The officers also decided that permanent residents of the campground area would be required to pay membership and dues: "all parties residing permanently at the Braden Castle Camp will be required to pay membership fees and annual dues regardless of ownership of camp equipment where they reside."[24] In the absence of land ownership anywhere within the park, this requirement disallowed those who did not own camping trailers to opt out of membership in the community's social structure.[25] At this point, the officers also began to proscribe the materials that would be allowed in the construction of dwelling places in the park. A motion was passed that the only material to be used in the "tenting section" of camp was canvas, except for the possible addition of wood floors at the discretion of the directors.[26]

Having previously written the history of the castle property, H. E. Robbins was appointed to write an article for the paper notifying the churches and fraternal orders that they "were at liberty to hold picknics [*sic*] on the grounds through the summer."[27] Echoing earlier uses of the ruins of Braden Castle by locals for Sunday afternoon picnics, this notification was in preparation for the return of the majority of the group's membership to their northern residences for the summer. Finally, the officers also mandated a rule that would influence the rapid construction of houses on the grounds when the Camping Tourists returned for the following winter season: "All property rights and holdings as to location may be changed to other locations when the holder has failed to erect a building inside of 18 months from date of allotment, if such changes are in the interests of the improvement of the cottage district, or in the judgment of the board, said board may make such changes as they see best."[28] This specification seems to contradict the temporal nature of the tourist community in its earlier version as the Tin Can Tourists. At the same time, it alludes to the goal of achieving a greater stability amidst fears of continued itinerancy and protracted conflict with municipalities. This requirement also shows the power invested in the board of directors to shape all aspects of the community—a power that was expressed in the final draft of the bylaws and one that will be explored further in the next section.

In their preparations to purchase the Braden Castle property, the officers of the Camping Tourists had consulted the mayor of the municipality with jurisdiction over the site to ask that certain concessions be made. The Tourists wanted the mayor of the town of Manatee to give up the city's authority and control over the property. R. W. Vaughn, acting secretary of the first board of directors, described the situation, "[we] had a talk with the Council and Mayor of Manatee in which we tried to have them agree to take the Braden Castle property out of the city but they would not consent to do this."[29] In spite of this initial refusal, Braden Castle Park has since achieved a degree of autonomy and is frequently referred to as a "city within a city."[30] The deputization of members of the park's community by the Manatee police chief exemplifies this semiautonomy: "In every sense, its [Braden Castle Park's] administration has accepted the ordinance of its parent community, Manatee. Officers of the law have been appointed by the Manatee chief of police, directly responsible only to him, with all powers of a city policeman within the confines of the camp, and offenders arraigned in the city court of Manatee."[31]

Within the partial independence of the park's self-government, the bylaws of this "city within a city" have served as the primary point of reference for the community's membership, governance, built environment, and adjudication. Typically, the term "bylaws" refers to the rules that internally organize an association or group of people but remain subsidiary to the wider-ranging rules of the land. Etymologically, "bylaw" can be traced back to the Old Norse combination of *bý-r*, meaning "dwelling-place," and *lag*, "fellowship," also connoting "law."[32] These meanings suggest that the modern "bylaw" also includes the ideas of settling disputes outside of traditional law courts by "specially deputed arbitrators" who listen to the testimony of neighbors.[33] This judgment by neighbors is an integral part of Braden Castle Park's community. In some cases the internal, purportedly subordinate regulations in the park's bylaws actually simulate the external legal structure of the society in which the park falls. The park's bylaws, though to a lesser degree, also regulate the community's dealings with the external public.

Making Camp: The Eutopic Construct

> OBJECT: To provide a winter home for American Tourists, under the best influences, for the mutual benefit of all, the improvement of health, the encouragement of education, and the moral betterment as well as the amusement and entertainment of all, and to establish good fellowship, and enforce clean and sanitary camp and cottage sites.
>
> —Constitution and Bylaws of the Camping Tourists of America (1945)

The objectives expressed in the preamble of the bylaws and constitution of the Camping Tourists of America outline a social, political, and spatial structure that is eutopic in its call for the best possible social order that can be afforded to its inhabitants within a camping setting. As a place of ideal happiness and of good, if not perfect, order, eutopia was one aspect of Sir Thomas More's *Utopia*.[34] In this project, More played on the antithetical qualities of eutopia as good place and utopia as no place to talk about the social structure of the imaginary island of Utopia. In early promotional literature for Braden Castle Park, the community was described as a type of eutopia: "Braden Castle Camp is, in fact, the Democratic Eldorado [*sic*] for a select group of transients who make this city and this county [Manatee] their winter headquarters."[35] The organization of More's Utopia parallels the bylaws and layout developed by the Camping Tourists of America by sketching a set of relations for communal life.[36] And in More's Utopia, as in the making of Braden Castle Park, social ordering is tied up in human construction.[37]

In spite of scalar differences, the plan of Braden Castle Park is similar to the geographic configuration of More's island of Utopia.[38] This comparison highlights the relation between the community's layout and its natural environment. The park is located on a point of land at the confluence of the Braden River with the Manatee River. At this juncture, both rivers are more than one-half mile wide and are influenced by tidal fluctuations. With the rivers forming two of the property's three sides, a tidal estuary that also flowed into the property's center formed the park's west-southwest edge and created the crescent configuration of the street and building layout. The positioning of the park along the two rivers is also reminiscent of the town of Amaurot that More singles out as a particularly eminent city in Utopia, with the River Anider and its unnamed tributary. More describes the island of the Utopians as a similarly shaped landmass: "These ends, curved round as if completing a circle . . . make the island crescent-shaped, like a new moon."[39] By 1936, the tidal marshland of the park had been converted to a landscaped lagoon that allowed for greater management of the tides. This idea of altering the natural geography is also found in More's Utopia, where man maintains complete control over natural forces such that the inhabitants have "changed its geography" from "rugged nature" to a more cultivated landscape much like the Utopians' gardens that were "so well cared for and flourishing."[40]

The grounds of Braden Castle Park include both the temporary camping grounds and the building sites for the Camping Tourists of America. The campground was originally conceived as a mechanism to generate profit for the group, and camping is still allowed today in trailers and recreational vehicles. The camping areas, designated as "trailer" in the bylaws, were one of three types of settlement in the park. Reflecting its evolution from the transience experienced earlier in De Soto Park, two semipermanent categories were introduced as "cottages" and houses within the main grounds.[41] The ten cottages were similar in size and degree of permanency to the main houses but were intended as rentable units that could generate a source of income for the park's residents. The bylaws note that each member, with an original minimum investment of $100 per stockholder, is entitled to a building lot that does not exceed forty by forty feet.[42] Buildings are required to be not less than three feet from the lot lines, to conform to the City of Bradenton's building ordinances, and to meet approval by the board of directors.[43] The materials allowed for in the bylaws are "wooden structures . . . [with] horizontal siding, or shingles, . . . or covered with stucco."[44] This material was intended to give a greater degree of permanence to the dwellings in contrast to the earlier canvas and tent shelters of the camp in De Soto Park. With the houses of Amaurot in More's Utopia, construction had also progressed from the original cabins "built slapdash" to the "handsomely constructed" houses of "fieldstone, quarried rock or brick."[45] The bylaws also limited the number of houses to one per lot and disallowed duplexes in order to avoid the possibility that one side could be rented out. Rooms built above garages must be "a part of the house."[46] The rules of construction were administered by the Building and Grounds Committee, comprised of three certificate stockholders who served for one-year terms and had to give written consent before new buildings were erected or old buildings were modified.

Today, 194 residences remain in Braden Castle Park.[47] Nearly half of these structures were erected in the first year of the park's existence as a result of the requirement that certificate holders build on their allotted land within eighteen months of the April 1924 meeting. Neglecting this deadline resulted in the forfeiture of the property back to the community. R. W. Vaughn noted that, as a result of this rule, ninety-five cottages were constructed between October 1924 and April 1925, along with a pavilion, a clubhouse, six rental cottages, and three commercial buildings.[48] This rapid construction amounted to a rate of construction exceeding one residence per day. Though following the rigid guidelines of an accelerated construction schedule and a limited palette of materials on a compact site, each residence was constructed as an individual expression of the certificate holders' preference for architectural form and style. Styles of residences in the community include Bungalow, Craftsman, and late nineteenth- and twentieth-century Revivals. The Sanborn maps from 1929 show the layout of the park as a nearly complete community with a marked similarity in the building footprints and a density of building structures not typically found in residential developments and thus more reminiscent of the compactness of the original municipal campsites like De Soto Park.

Maintenance of the campgrounds and construction of communal buildings were originally shared by the first certificate holders in Braden Castle Park. With space for six hundred people, the community hall was built entirely by members of the association. A manuscript written on March 12, 1925, records this construction project: "When the project of building the community hall was suggested a census of the camp revealed that there were carpenters, brick masons, plumbers, painters, decorators, electricians,

Braden Castle Drive in Braden Castle Park. Photograph by author.

and every kind of artisan whose efforts could contribute to the project for the common wealth."[49] With a similar emphasis on creating a place where "everything has been well-ordered and the commonwealth properly established," More's Utopia includes an "abundance of everything" because of "everyone working at useful trades" and the ability to "assemble great numbers of people to work" on such communal projects as road building and repairing structures.[50]

The paradox of freedom and control found in the camp's physical composition also reflected the Camping Tourists of America's modes of self-governance. Similar building footprints allowed for a diversity of building styles, just as the bylaws framed selective rules but were at the same time open to interpretation by both certificate holders and members of the Building and Grounds Committee. The bylaws include the provision that each member "in good standing" be entitled "to one vote only."[51] These voting rights can only be granted internally, because the tourists do not have permanent state residency and thus lack the right to vote. This democratic basis is tempered by the oligarchic power structure of the nine members of the board of directors, who "transact all business of the Association, three of whom shall be elected at the annual meeting of the members, to replace the three retiring directors, and they shall serve for a period of three years."[52] Though characteristic of contemporary neighborhood associations, the duties of the board to exact penalties exceed the prosaic administrative tasks of managing funds, overseeing membership, and directing meetings. The penalties section of the bylaws outlines the adjudication of violations within the community, simulating a court of law in which the president of the board of directors serves as judge. Although not acted upon in the memory of any of the current members interviewed, the first part of the section on penalties describes the conditions under which a member may be found in violation of the bylaws: "Any member of this Association knowingly or willfully violating any of the provisions set forth by these by-laws, or any rules or regulations made under them, or who shall be guilty of *agitation intended to disorganize the morals or best interests* set forth in the preamble . . . may . . . be tried before a jury of 12 certificate holders."[53]

This legal system does engage to a certain extent the models of the State of Florida. The bylaws state that the "jury shall be selected in accordance with the laws of Florida for selecting jurors." If convicted, the "guilty member" has thirty days to vacate his or her property and sixty days to "sell or dispose of his certificate, to a purchaser acceptable to the Directors." Other less nefarious negligence, such as refusal or neglect to pay delinquent taxes or to comply with the bylaws, will result in the offending member's name being reported and listed on the books of the association. In addition, a "list of such members who are not in good standing shall be read at the annual meeting, and a copy shall be read at the annual meeting," when a two-thirds vote by the membership can revoke the aberrant member's certificate. Other prohibitions include the "use of intoxicating liquors and the trafficking in the same" and the performance of "manual labor . . . [or] any public games played on the grounds of the Association on Sunday." All pets are discouraged and are "subject to the regulations adopted by the Board of Directors."[54]

Breaking Camp: "Strangers at the Gate" of a Tourist's Home

The founders of Braden Castle Park envisioned a boundless future for their community. Speaking through the guise of the castle, Robbins concludes the castle's soliloquy with an invocation to the mythology of the place and classifies the camp as hallowed ground: "Again in reverence to the many friends that crossed the great divide and to the pioneers that made and shaped my walls, I say, may their spirits ever hover over these sacred grounds, and bless and encourage the onward trend of advancement that will perpetuate my memory to future generations, and the continuance of this tourists' home throughout coming generations."[55] More also sees the possibility of limitless longevity in the construction by the Utopians: "Through the plan of living which they have adopted, they have laid the foundations of a commonwealth that is not only very happy but also, as far as human prescience can tell, likely to last forever."[56] In one respect, the longevity of the Tourists has depended on their attachment to the early history of the place and the physical components of the place itself. In the unrefined spiritualist-humanist model proposed by Robbins, tourists rather than citizens of the place make the legendary grounds home and are given the responsibility of remembering and advancing the ideals begun by the "pioneers." Robbins has also formed a creation myth that, even though ingenuous, attempts to reconnect the entire project to the elemental materials of the Florida landscape—the "beautiful white sands at low tide," the "shell life of the aquatic environments," and the "shimmering waters of the Manatee." And these materials are found in the castle's tabby construction.[57] Recounting the process of its creation, the castle begins its soliloquy by describing the formation of these tabby walls:[58] "Well do I remember when I first saw the light of day and awoke from that trance to a realization that I was a changed body from that of aquatic environments made up of shell life, that had counted the ceaseless changes of the tide that lashed my various forms at will. . . . I became conscious of my surroundings when I fully awoke from that state of transformation to the realization that I was a form that was new . . . destined to mark the early trend of domestic and social development in this land."[59] Robbins makes it clear with this statement that the Camping Tourists saw themselves as generating a "new form" of development that would revolutionize domestic and social life.[60] The goals of health, education, "moral betterment," amusement, and fellowship cast within a cultivated Floridian landscape are achieved through the built environment of a "tourist home that has no equal in the world."[61] This new form provides a home-away-from-home for the Tourists, recently evicted from their temporary lodgings in nearby Tampa.[62]

The Camping Tourists of America attach to the constructed history of the place. Inserting themselves into this history through such devices as Robbins's castle soliloquy, the Tourists sought to make a home in spite of their condition of estrangement. Robbins, through the castle's voice, notes this situation: "how this gladdens my life as I view this wonderful transformation of the past. To these environments come people from the Northland—from East and West—to this home of flowers and verdure, and many other semi-tropical fruits that bless the *stranger within our gates*."[63] In this situation, the utopic and the eutopic occur simultaneously within the construction of Braden Castle Park. In their rigorous bylaws, the Tourists have attempted to construct a "good place" by restricting vices, cultivating an environment of morality, and regulating construction practices. At the same time, the campers must maintain the imagined "no place" of their still ephemeral existence within the equally transient Florida wilderness.[64] As habitual strangers to a place they call home, the Tourists narrate a history that reconstructs the ruined tabby walls of the castle and at the same time build more permanent shelters to avoid displacement. H. E. Robbins himself built a home immediately across from the original castle ruin that he named Braden Castle II. In the soliloquy, Robbins is self-referential when he has the castle say, "Then just across the road to the south was erected a large residence that made me think of the time when I was young, and the Bradens were with me." Sentimentality blends with a collapsing of time to reinvent the Braden myth.

The camp becomes the site for this confluence of utopia and eutopia and the resulting transformation of the place into a newly formed Braden Castle with the best possible social order that can be afforded to the camping tourists. The opening section of Sir Thomas More's *Utopia* includes a poem that reflects the transformations of the tourist camp between utopia and eutopia:

> 'No-Place' was once my name, I lay so far;
> But now with Plato's state I can compare,
> Perhaps outdo her (for what he only drew
> In empty words I have made live anew
> In men and wealth, as well as splendid laws):
> 'The Good Place' they should call me, with good cause.[65]

Robbins's epigraphic poem at the beginning of the "History of Braden Castle" echoes the work in *Utopia* with its pairing of rhymed lines and its invocation of a "good place" to call home:

> You'll find in the tourists that make up this home
> A swell lot of people that annually roam,
> They seek out the best in Florida's clime,
> That's why they come to this home that is fine.[66]

Thirty miles north in Gibsonton, carnival performers camp within the re-fashioned spaces of the midway. The iconic potential of the midway is, like the environs of the Braden Castle's ruins, a new kind of home.

Chapter 8

GIBSONTON

Parasitic Regions of the Carnival Midway

Arriving: Gibsonton, Florida, as Fish Camp

Gibsonton is an unincorporated collection of trailers and mobile homes on the south side of the Alafia River in Hillsborough County, Florida.[1] Gibsonton has been home to many of the nation's sideshow performers like Melvin Burkhart, the original Human Blockhead, Percilla the Monkey Girl, and the Lobster Boy. Not immediately visible to the auto traveler, this local identity occupies a seemingly unremarkable setting typical of much of Florida's exurban landscape with its uneven assortment of vehicles and houses, both mobile and immobilized. Most recognizable is Giant's Camp Restaurant that makes up the formal nexus of Gibsonton and greets the southbound traveler along Highway 41.

As a political entity, Gibsonton and its boundaries are more difficult to define. While governmental overlays have impacted the area's zoning and its edges, Gibsonton's assemblage of semipermanent housing units and commercial apparatus eschews classification as a "town." Known locally and colloquially as "Gibtown" and "Showtown U.S.A.," Gibsonton falls between the conventional denotations of town and city.[2] Instead, what might be termed its "town-field" gains some definition from geographic features, at once isolating and propitious, with the Alafia River to the north, Interstate 75 to the east, and East Bay's tidal flats to the west. What truly identifies the "town" and its many connotations is the unexpected yet not surprising encounter with concession stands carefully placed in front yards, with a Ferris wheel rising from behind a mobile home, or with residents' accounts of caged lions and tigers roaring across side-yard fences. Today, the camptown maintains an idiosyncratic yet communal identity, its carnival history visible in the only United States post office with a special counter for midgets, in a museum of dirt collected on camping excursions around the world, and in its yards.

Gibsonton's position between Tampa and Sarasota along Highway 41 has made it a convenient location for performers and operators in the carnival and circus business during the winter season. In the early 1920s, Eddie and Grace LaMay were the first carnival performers to settle in the fish camp that had formed adjacent to the Highway 41 bridge on the north side of the Alafia River. The LaMays decided to make this location their winter home after they stopped to fish during an overland tour in the carnival off-season and were unexpectedly welcomed by those living in the camp. In the late 1930s, Al and Jeanie Tomaini, who became famous in carnival performances as the Giant and the Half-Girl, began parking their travel trailer along the river in the Alafia fish camp during the winter season. Ruth Pontico, who lived in nearby Tampa and performed with the Tomainis as the Fat Lady in circus sideshows, had introduced the couple to the area.[3] In 1940, the Tomainis purchased 3 1/2 acres on the south side of the river. As noted by their daughter, Judy, the rationale behind the purchase included the favorable fishing location but was primarily a function of the parcel's location immediately south of the Alafia Bridge.

The Giant, Al Tomaini, established the eponymous Giant's Camp at the apex of this wedge-shaped plot that is flanked by Highway 41 on the east and Lula Street to the west. The Tomainis continued to develop the campsite through 1952, clearing and filling the swampy areas of the property for trailer sites and one-room cottages. Judy Tomaini notes that Frank Lentini, the Three-Legged Man from Miami and a close friend of the couple, had encouraged Tomaini to name "the place" Giant's Camp to attract both tourists and seasoned carnival performers.[4] The Tomainis constructed a bait shop to service the fishing community and maintained the existing bar as a restaurant, which is the present Giant's Camp Restaurant. The original bait shop was destroyed in 1997 after Cargill Phosphate, the company that owns the bay access, refused to honor the Tomainis' waterfront lease.[5]

Al Tomaini, "The Giant," in front of Giant's Camp, 1950. Courtesy of Judy Tomaini.

Siting Camp: At Home on the Midway

(and it is part of my good fortune not to be a home-owner!). But if I had to . . .
—FRIEDRICH NIETZSCHE, *The Gay Science*

Although the site's general location is a function of the natural resources available from its proximity to river and bay, its specific placement and orientation is directly tied to the Tomainis' decision to resite the settlement. With this purchase, the Tomainis sought to achieve the greatest impact on travelers approaching what would become Giant's Camp from the north along Highway 41. Their experience with the layout of sideshows and midways in fairgrounds influenced the placement of the camp on the right side of the highway immediately south of the bridge. This "first addition to Gibsonton on the bay" followed rules learned from observed human nature at the carnival midway where visitors, consumers, and fairgoers typically turn to the right when entering a new space. Judy Tomaini repeats this carnival precept practiced by her parents, "when setting up on a show lot, the most desirable location was first in, on the right."[6] From the outset, the Tomainis conceived of the camp as a fairground and foresaw the potential for Gibsonton's expansion to the south, with their camp forming the nexus and the threshold of the larger settlement. With the Alafia Bridge serving as the entry gate, Giant's Camp, "first in, on the right," remains the cultural and spatial focus of contemporary Gibsonton.

In addition to this fortuitous and symbolic location of Giant's Camp, siting Gibsonton entails the carnival performer-residents themselves. Person, persona, and personality are inextricably linked to the symbolic potential of Gibsonton's places. As "campers" in Gibsonton, the performers gain an anonymity—becoming "locals" within a temporary homesite affording them the place to be simultaneously "in" and "out" of character.[7] But this anonymity also hints at an ambiguity that is fundamental to a preliminary understanding of Gibsonton, as a place and as a campsite for intersections of domestic space and carnival. The performers are itinerant symbols who, in their unique combination of mobility and changeability, can move fluidly from sideshow to side yard at home. In at least one case, this characteristic flexibility allows for the Gibsonton resident known as the Enigma to appear, almost seamlessly, as himself on an episode in the *X-Files* television series.[8] Though filmed on location in Gulf Breeze, Florida, the show ostensibly takes place in Gibsonton, where Fox Mulder and Dana Scully are investigating the death of the Alligator Man.[9] With his name changed to the Conundrum in the episode, the Enigma serves as a poetic device linking the plotlines and symbolizing the difficulties of solving a mystery in a place where all characters and attributes are presented as "suspect."[10] Watching the Enigma move through this televisual nonplace, we might begin to understand how the equally poetic characters of the "actual" Gibsonton live within the equally ambiguous, even enigmatic, spaces of our own increasingly mobilized campsite-world. Gibsonton's spaces and its residents can thus be read as sites and personas of intentionally obscure metaphoric capacity—making up a parabolic story to be experienced as much as read, or viewed on television.

With its performers as campers, the midway becomes a camping spectacle. This conceptual and formal resonance between camp and midway is neither incidental nor unique to Gibsonton—sideshow performers have historically camped in trailers and tents attached to the back of their show trailers that fronted onto the midway. Also, in its assembly, the midway like the camp is designed as a demountable construction that can be taken down and rebuilt in another location. As such, the carnival midway is passage, connector, and destination. This dual meaning works both conceptually and historically. The term "midway" derives from Chicago's Midway Plaisance, which Frederick Law Olmsted designed as a linking boulevard between lakefront Jackson Park and inland Washington Park in the early 1870s. Although never completed to Olmsted's specifications, the Midway Plaisance did, however, serve as a component of the World's Columbian Exposition of 1893 and consequently gave its name to subsequent carnival expositions and fairs. Thus, in this particular instance, the Midway Plaisance served as a wide connecting corridor between two park districts and showed its potential as a more static destination place functioning as the site for the Exposition's carnival and fair component. The concept of the midway is symbolic itself, particularly in its ambiguity as a zone that links one place to another, as threshold, and signals an arrival to conclude a passage, as edge. As a midway space, Gibsonton enfolds the passage of tourists and maintains a liminal position along the geographic feature of East Bay. The concept of the midway also symbolizes the sequence of clearing, preparing, and marking a site—integral to making a campsite into a homesite.

Anchored by the precisely located Giant's Camp, Gibsonton as midway articulates both a connecting mode of passage and a stopping point for tourists. From the 1920s up to the installation of the Interstate system, Highway 41 served as a major north-south corridor between Miami and the Upper Peninsula of Michigan.[11] The town of Gibsonton is thus sited along one of the main connectors of tourists and vacationers as well as the carnival performers themselves. In this reconfigured midway space, vehicular traffic along Highway 41, specifically winter season "snowbird" migrations from Canada and the northern United States, substitutes for the foot traffic of fairgoers in the summer season, and the performers' off-season hometown remains a well-traversed public corridor. As midway, the town is also a destination, a spatial hesitation between other points of interest. By the early 1950s, Tomaini had established the tourist camp around his eponymous restaurant; and the Giant himself, along with the highly regarded fishing, served as the place's main attractions. The town's location was fortuitous, providing an alternative to Tampa's urban density and occurring at the midpoint between Miami and common stopping points to the north, such as Valdosta in southern Georgia.[12]

In early American fairs, the midway zone occurred outside the fenced area of the agricultural fairground. Fred Kniffen notes that this model was common in American fairs by the middle of the nineteenth century.[13] Performers set up their show tents and temporary lodging in this location of exclusion. Gibsonton similarly maintains a peripheral relation to the city of Tampa. "Gibtown," as it is identified by its residents, remains unincorporated and occupies the East Bay zone where power plants, phosphate companies, and landfills have traditionally been located.[14] As a camptown, Gibsonton can be read generally as the zone outside the city walls, analogous to the open fields of the Italian *campo*,[15] the Campus Martius of Rome, and more generally Greek *chora* as opposed to the *polis*.[16] In the American landscape, early tourist camps of the late 1910s were sited as informal arrangements for overnight stays in a farmer's yard chosen for the availability of well water.[17]

In Gibsonton, these residual spaces are potential sites for installation of the midway apparatus along the town's public edges, particularly those spaces along Highway 41. Although less formalized campsites remain in the easements and interstitial spaces of Gibsonton's half-places (*mi-lieu*), the camptown as a whole is in the process of becoming what Michel Serres calls *milieu*.[18] In this process of growth, the excluded middle (*mi-lieu*) becomes *milieu*, just as the supplemental or "receptacle" space of *chora* becomes increasingly occupied. And American midways and tourist camps, such as Gibsonton, develop into more permanent destinations. Serres has described the excluded middle as the parasitic operator that sets the stage for a multiplicity of relations. Excluded, this operator parasites the relation *between* components rather than the components themselves.[19] The characteristic fluidity of this operation and activity is found in the specific practices and itinerancy of the camp and the midway: "Hosts and parasites are always in the process of passing by, being sent away, touring around, walking alone. They exchange places in a space soon to be defined."[20] Within this process of parasitic "becoming," the midway and the camp remain effectively "threshold places."

In the specific case of Gibsonton, Giant's Camp was understood from its inception as a semipermanent winter quarters and the planned anchor for a larger camp settlement. Having founded the camp with his wife, Jeanie, Al Tomaini served as the "Lot Man" for the camp's development over time. On the midway, the Lot Man typically organizes, divides, sets up, and eventually breaks down the rides and show venues. Just as Tomaini cultivated the siting of Giant's Camp through time, the Lot Man oversees the installation of the midway, allotting and siting each component of the temporal assemblage. Moreover, Giant's Camp becomes the entrance-threshold to the community, just as the midway serves as the gateway to the fair.[21] This relation continues with the Lot Man's understanding of the emblematic carousel, or more traditionally the Ferris wheel, as the "key" to the midway.[22] Additionally, the camptown's characteristics of the excluded middle transform the typical differentiation of the public and the private. In this ad hoc urban planning, the distinction between campground and fairground is blurred, allowing for the private camp to serve a performative function and for the publicly oriented fair to achieve a degree of privatization. This process is not a simple inversion but a type of overlap and mixing—what might be called the *milieu* of the public and private.[23] Along Highway 41, the resulting uneven development mixes mobile homes, trailers, commercial buildings, and large-scale carnival trailers, typically displayed in the town's parking lots—all fronting onto the main thoroughfare.[24]

Clearing Camp: Zoning Places and Placing Zones

Gibsonton maintains both a distance and a proximity to the urban areas of Florida's west coast. The town connects to Tampa and Hillsborough County's existing infrastructure of power, transportation, and other services. Discussing the organizational influence of services, J. B. Jackson notes that the American vernacular home relies on community "not as political entity, but as a source of services."[25] In this model, the provision of water, electricity, and transport systems locate and structure the idiosyncrasies of local developments that are otherwise characterized by the fluctuating processes of the vernacular's "living organism."[26] Jackson's vernacular homesteads, like the campsites of Gibsonton, result from "hesitation," or a temporary immobilization that typically endures: "Mobility and change are the key to the vernacular landscape, but of an involuntary, reluctant sort . . . an unending adjustment to circumstances."[27] This duration occurs within the fluctuations of time and thus allows a flexibility for adaptation and ultimately a place for invention. Through these reflexive and highly circumstantial transformations, the vernacular constructions of camp are able to parasite the service systems unobserved, though typically not through deception. In their own relatively small and tentative operations, camps subvert and thus benefit from the large, authoritative, and slow-moving bureaucratic mechanisms of municipal, regional, and state-run systems. At the same time, camps develop subsystems that negotiate these larger entities through mobility and flexibility. Camps such as Gibsonton are Jacksonian vernacular par excellence, assimilating change and coding themselves through this "unending adjustment to circumstances" that is systematized in the camping process, in particular the procedure of clearing.

Gibsonton, as reconstructed midway and as camptown, has developed its own subtle codes of place-making, but its early success ironically came out of the planning and zoning rules adopted by Hillsborough County. This rezoning of the area in and around Gibsonton has resulted in the settlement's rapid expansion and growth. In the 1950s, a zoning overlay called "Residential Show Business" was added to allow for the display of carnival equipment in the front yards of performers and operators who were increasingly drawn to Gibsonton's emerging community. This zoning revision recognized the area's ad hoc community (the area known locally as "Gibtown" and colloquially referred to as "Carnietown" by nonresidents) and coincided with Sarasota's restrictions on such public exhibits. Attempting to alter its own image as a winter haven for carnival and circus performers and to augment its growth as a retirement village and cultural center for southwest Florida, Sarasota revised its planning and zoning ordinances explicitly to prohibit the display of circus paraphernalia on residential property. The Ferris wheels, funnel cake stands, and lions and tigers moved north to the front yards of Gibsonton. The Show Business (SB) district in Hillsborough County expressly accommodates "the special needs of business and residential uses related to circus, carnival, and other show business activities" and provides "opportunity for the special housing patterns needed by these business persons," ultimately to encourage the "grouping of those land uses having specific interrelationships with the show business activity."[28] The intentioned publicity that began with the siting of Giant's Camp continues at the scale of individual sites with the introduction of publicly displayed equipment in private yards, whether under repair or in partial service.

Along Highway 41 and other main roads of Gibsonton, carnival rides occupy commercial yards and parking lots. Roller-coaster rides, with their structural and rail components laid out and inventoried in the open yard, rest on trailer beds adjacent to box trucks with rear doors thrown open, lingering in a partially unloaded state. Other roadside displays call for the more active participation of viewers and passersby. Along the east-west road called Gibsonton Highway, a concession stand selling cotton candy and popcorn occupies the southeast corner of a yard with a structure that is ambiguously commercial, a dwelling attached to its north side. The stand presumably services the bus stop immediately across the sidewalk and advertises to drivers heading east into Gibsonton from Interstate 75. Another assembly of carnival components, also along Gibsonton Highway, is anchored by the "Midway Stop and Shop," around which a sausage stand, shooting-gallery booth, fun house, and tilt-a-whirl ride are for sale. Erected with a design and intentionality directly linked to the midway's own layout, the carnival pieces clearly define a central space in which the tilt-a-whirl ride is protected by the movable fencing components typically used along the midway to organize the carnivalgoers' circulation and queue spaces. In a residential lot, another concession stand advertising popcorn and candy apples has been parked in the front yard's corner so that its glazed service windows face onto the street and into the yard space.

In addition to its identifiable character as carnival town, Gibsonton's residences are primarily mobile homes, trailers, manufactured houses, and site-modified versions of each. This zoned reformulation of private-yard usage is matched by a consideration of the questions of permanence associated with mobile home communities. Ironically, zoning classification of lots for mobile homes and trailers is divided into "temporary" and "permanent" designations. Undefined within the county's comprehensive plan, temporary mobile home permits are processed through the Hillsborough County zoning counselor on a case-by-case basis. The entry for permanent mobile home installations reads: "Zoning for Permanent Mobile Homes: Mobile homes may be located on all legally created agriculturally zoned properties (except AS–1 and Al), as well as residentially zoned properties with a MH designation. In zoning districts that allow mobile homes, only one dwelling unit is allowed on a lot. Often there are cases when someone would like to place an additional residence on a property (e.g. medical hardship, farm worker housing, etc.)."[29]

Through the exclusion of the term "temporary" from its administrative

Cotton candy and popcorn stand in front yard, Gibsonton, Florida. Photograph by author.

text, the comprehensive plan recognizes the historical and practical precedent for the permanence of mobile home placement. Understanding the typical fixity of mobile homes, the architect Paul Rudolph, who practiced in this area of central Florida, had already recognized what the comprehensive plan presupposes as the precept "once they are there, then they are there." The plan's terminology "permanent mobile homes" identifies a paradoxical situation of time, place, and home. The temporal nature of the mobile home, even in its zoned permanency, provides a place conducive to the spectacle of the grand outdoor storage spaces.

Gibsonton also maintains and transforms the local and national political apparatus. Gibsonton's collection of performers and their concomitant culture of the carnival has led to the county government's recognition of a community not only in the Show Business Overlay but also in the Gibsonton NFZ, or "No Fee Zone": "This program administers Hillsborough County's five (5) general government impact fees: transportation, right-of-way, park, school, and fire. These fees are charged to new development, both residential and non-residential, to help pay for the structure's impact on the road, park, school and fire network. The County also has impact fee relief programs such as the No Fee Zone (limited duration) and the Affordable Housing Relief Program to encourage development in economically distressed areas. Economic development incentives are provided to qualified business creating quality jobs in Hillsborough County."[30]

Similar in conception to the Urban Enterprise Zone, the NFZ is an area that is determined by a need for impact-fee relief, allowing economic development to forgo paying fees for the expected impact on existing infrastructure (roads, parks, school and fire network). Another way governmental bodies have attempted to delineate Gibsonton's unincorporated community is evident in the U.S. Census's characterization of the area as a "Census Designated Place" (CDP). The Census Department defines the CDP as follows: "A statistical entity, defined for each decennial census according to Census Bureau guidelines, comprising a densely settled concentration of population that is not within an incorporated place, but is locally identified by a name. CDPs are delineated cooperatively by state and local officials and the Census Bureau, following Census Bureau guidelines. Beginning with Census 2000 there are no size limits."[31] The loose aggregate of Gibsonton's community is further defined by governmental superimpositions that benefit the camp's population but are not required for its internalized identity.

These attributes outline situations and practices of simultaneous attachment and detachment. At the territorial scale, these features can be mapped in terms of zoning regulations, aerial photographic research, and governmental directives. But in each case at this scale, the Gibsonton community maintains its unique characteristics *by way of*, rather than *in spite of*, these governmental regulations, overlays, and "overviews." This contradictory relationship actually inverts typical links to zoning ordinances held by unincorporated areas and exemplifies a vernacular construct that parasites the politic of power. In a sense, Gibsonton zoned itself, drawing from the midway's conventions and traditions and using the existing apparatus of governmental regulation. This unique self-regulation is paradoxical in its reliance on a system-in-place, which can be appropriated and used to regulate its own blurring of the temporary and the permanent and the public and the private. This process of reinvention is inherent in siting Gibsonton's camp and in setting up the midway. Camp and midway coordinate a flexible, yet rigorous, rezoning of place: "The Midway lingers—residual, delinquent—as a premonition, a possible model for a less possible architecture. The Midway wanders only to repeat its parasitic manufacture in a succession of marketplaces."[32] In such a system, the "excluded middle," or parasite, is both a component of the relation and the producer of change within the relational matrix of the system, and it is the fluidity of Gibsonton's camping community that has allowed for this condition.[33] Camps and midways, as procedures of "clearing" and constructing place through time, outline the practice of an open system, one in which making and remaking are symbiotic.

Making Camp: Clearing and Collecting Shell Gardens and Museums of Dirt

Any process implies a system, but not all systems imply process.
—ROBERT MORRIS (1970)

In addition to its place-making activities at the territorial scale, camp practice occurs at the level of the detail. As a system composed of multiple activities, camping is a process that includes the making of both camp and midway. Taking up the aphoristic statement by Robert Morris, quoted above, we can say that camping implies process through the systematization of the parasitic mode and its associated local operations. At the scale of detailing, the construction of camps and midways is an architectural procedure of indirect action. Such methods relate to the concept of bricolage with its origins in the French term *bricole*, which is an unexpected stroke or an occurrence brought about indirectly. Claude Levi-Strauss expands on the meaning of the term to include the response to an obstacle: "It [*bricoler*] was however always used with reference to some extraneous movement: a ball rebounding, a dog straying or a horse swerving from its direct course to avoid an obstacle."[34] Such architecture relies on maintenance as a way of transforming its "shape" rather than on the application of a plan or formal mandate. Camping relies on, and thus must maintain, this contradictory relation by "making do" with adverse circumstances or unlikely materials. The paradoxical role of the parasite can be read at this scale through its "transforming by maintaining." Within the standard series of actions that comprise camping (siting, clearing, making, breaking), another array of operations is inserted within the camping sequence to form temporary alignments and connections. Bricolage becomes method. Michel Serres qualifies Levi-Strauss's characterization of bricolage with the introduction of the joker-parasite: "Placed in the middle or at the end of a series, a series that has a law of order, it [joker] permits it to bifurcate, to take another appearance, another direction, a new order. The only describable difference between a method and *bricolage* is the joker. The principle of bricolage is to make something by means of something else, a mast with a matchstick, a chicken wing with tissue meant for the thigh, and so forth."[35] In camping practice, bricolage is a function of necessity and play, and the "joker" might be a tent repaired with a shirt or scrap of fabric or a welded fusion of concession stand and outbuilding.

If we combine this idea of the joker-parasite with the argument that the vernacular is actually constructed of distant and unfamiliar components, we understand camping as a patchwork of foreign and local places and past and present events through the particular situation of Gibsonton.[36] Camping elides and inverts distance and time through its set of operations. But the bricolage of camping is not a construct simply stitched together through a postmodern ad hocism.[37] The siting of camp arises out of the specific location and at the same time comes about "from elsewhere." This "specific elsewhere" has a duration that works between the present camp and past instances of camp. Camping's present "was" and its past "is."[38] The operational sets for making the bricolaged duration of the camp are movement (vector), transformability (joker), and the growth of relations (parasite). Gibsonton endures with the refashioning of these operations and the remaking of its camping spaces.

In this place-making game, specific procedures work between vector, joker, and parasite. The following operations can be found within Gibsonton and in the construction of the midway at the Florida State Fair in Tampa: laying out, edging, clearing, gathering, layering, surfacing, inserting, placing, filling (in), grafting, laying, folding, blocking, staging, accreting. These acts of making require human presence and, although at times reduced to modalities of assembly, remain connected to human experience through scale, materiality, and improvisation. While these activities are sometimes carried out in sequences, their application depends on particularities of site and context. Thus the process is not exclusively linear but forms a networked system of actions.[39] The midway's Lot Man must recast the construction at each new location to fit specific configurations and conditions.

The layout of the Florida State Fair's midway is a good example of the deviations from the midway's prototypical horseshoe shape. Also, the "fill-in" space varies with each new set of rides contracted for the fair. The Lot Man identifies the leftover space between the midway's rides and assigns these diversely shaped sites to vendors and smaller-ride operators according to his knowledge of their spatial requirements and their proximity to other rides. A comparison can be made with the problem of fitting and arranging single-wide mobile homes at Giant's Camp, which was originally laid out for the much smaller autocampers and travel trailers. The operations, including the partial list above, can be seen as patches that eventually form the network, or patchwork, of the camp and midway. These patches can be generally grouped under the headings of gathering (accretion), arranging (fill and infill), and assembling (graft and attachment). In each case, the method is an end in itself; and process is integral to the system used. Method is not applied but carried out and arrived "at." This method, however, does retain the indirect action, characteristic of bricolage—particularly in the way that the interventions of camping meet the ground.

Camping here relates to domestic gardening. In the garden as well as the camp, a multiplicity of actions, the tilling, seeding, pulling, weeding, covering of gardening, is used; and formal design often succumbs to the necessities of maintenance. Moreover, the private garden's association with the house and yard reappears in the practice of camping. As the Gibsonton yard becomes the garden, the open *hortus* that its etymology implies, the possibility of renewed movement also permits nonstandard uses of the site.[40] The private carnivals and circus displays are remade or "grown" with the return home each season. Equipment is not simply stored but "planted" and shown to a general audience made up of neighbors, sightseers, and tourists. This public gardening, an arrangement of signs,[41] is similar to the process exemplified in a shell garden photographed by Marion Post Wolcott in the Sarasota City Trailer Park.

Shell gardens were commonly used to decorate the supports of the trailer hitches, the inherently locative points of the campsite.[42] Accordingly, the garden is located around the trailer component that is connected to the ground through a temporary support—the procedural nexus of planting and uprooting.[43] The temporary shell gardens not only decorate the ground around the trailer-hitch supports but also identify the trailer owners by spelling out their surnames. The caption of Marion Post Wolcott's 1941 photograph of a "guest at Sarasota trailer park" identifies the garden components as "shells and odds and ends," describing a method that makes one thing from something else.[44] Michel Butor has pointed out that in addition to the knowledge of making something, bricolage entails the collecting of things. According to Butor, the two levels of bricolage are the knowledge of making something and the collection of these things that have been neglected. This second level of bricolage represents the urge for infusing an unknown meaning into abandoned things.[45] Such a collection, or accumulation, allows for a "traveling-in-place," in the same way that the shell garden collects the camper's identity and memories of place.[46] Camping bricolage occurs in Gibsonton as a hybrid of collected program and as a kit-of-parts assemblage.

In Giant's Camp, one of the cottages served as Judy Tomaini's office for her business "Rock's Monuments" and as an informal museum of celebrity, family history, travel, and the camp's history. In addition to six hundred autographed celebrity headshots, the office-museum held a collection of dirt from around the world. Tomaini's dirt collection includes soil that she herself has excavated as well as samples of earth sent by friends from the places they have traveled. Tomaini's specialty is grave dirt, particularly dirt taken from plots at the Showman's Rest cemetery, where her mother, Jeanie Tomaini, and father, Al Tomaini, are buried. The itinerant life of the carnival performer leads to a unique problem of choosing the burial sites for family members. In Gibsonton, the community of performers has established two cemeteries for the burial of showmen and -women.

In the tradition of the kit-of-parts assembly, a decommissioned observation wheel has been rebuilt in a Gibsonton backyard. Serving its traditional purpose as the "key" to the midway, this wheel becomes the symbolic anchor to the carnivalesque spectacle constructed in the off-season; and this resident, having painstakingly relocated the wheel, becomes its private Lot Man.[47] While much smaller than the original Ferris wheel, its relative scale is comparable to its surroundings, set against the domestic dimensions of parked mobile homes.[48] With the reconstruction of the wheel in the space of the backyard, the now completely privatized and partially defunct wheel becomes wholly symbolic. Referencing the experience of the fair and carnival, the apparatus signifies the vertiginous experience of suspension and the

Ferris wheel in yard, Gibsonton, Florida. Photograph by author.

unfolding of panoramic space—as fairgrounds, or in this case townscape, with each revolution of the wheel. The private Ferris wheel refers to the invisible and inaccessible spaces of the resident's backyard, to the spatiality of the potential panoramic views of Gibsonton's yards, and ultimately to the town's relation to the bay and to the city of Tampa. In other instances of yard as midway, the spatiality of the symbol itself is evident.[49]

The reconstruction of the carnival midway's spaces occurs at the scale of the individual yards in Gibsonton. When the carnival performers set up their dwellings on the road during the summer season, the midway serves as the primary space to which the dwelling space is attached. The midway becomes the performer's yard. The owner's lodging takes the form of campsites, trailers, and tents that are often set up directly behind their concession stand, ride, or sales booth. This situation is reversed in the town of Gibsonton; the yard is often transformed into a reconfigured midway space.[50] As a primarily symbolic overlay, the midway apparatus, though it is now secondary to the site's primary use, characterizes the identity of the place. The yard not only signifies home but becomes a symbol of performance. This situation occurs most dramatically with the relocation of the Ferris wheel in the mobile home's yard, but is also emblematic of the camping home.

In camps and camping, the mobility of home suggests a rhetorical territory, unique to vernacular constructions.[51] If each yard display is a private demonstration of home, as a version of domestic life countering the neutrality of factory-produced mobile homes, then the public aggregate of homes within the camp becomes a political entity defined by an internalized identity and an externalized parasitism of the municipality. This communal concept of home among the camp's ad hoc citizenry reflects one aspect of J. B. Jackson's working definition of the vernacular. For Jackson, vernacular constructions are identified with the adherence to a commonly held idea about place and community, often apart from political influence.[52] Inscribing the "rhetorical territory" of home, these "spaces of belonging" serve as "mutually dependent processes of exclusion and identity construction."[53] As a vernacular construction, Gibsonton "camp" follows Jackson's pattern without being completely separate from political influence. The camptown's "rhetorical territory" maintains, transforms, and "makes" the local and national political apparatus.

The town's territorial attributes do not, however, remain solely rhetorical, and Gibsonton's symbols are not static. The organic nature of Gibsonton's symbolic environment is tied directly to its tradition of transience. Out of necessity, carnival performers work on the road. In this case, mobility is not metaphoric but real.[54] The spaces of Gibsonton's yards accommodate the dual movement and simultaneous transformation of camp to town and town to camp. Such temporal existence is defined seasonally, spatially, and economically. The pairing of midway and home allows for continuity between winter and summer seasons. Summertime means movement from fairground to fairground and from campsite to campsite. During the winter season, a tentative permanence concretizes the midway-home relation in the spatial symbolism of the yard constructions. The spectacle that is created is not image but instead activates experience through spaces that mediate street, yard, and house. As a lived symbol, experienced by the viewer-interpretant and performer-resident alike, the reconstructed midway serves to locate the carnival performer's home. *Any* space in the region becomes *some* place *through* the construction and performance of the midway.[55]

Breaking Camp: The Time of the Parasite

The activity of breaking camp, in the cyclic sequence of camping practice, is rarely understood as a final operation within the process. Instead, breaking camp, like disassembling the carnival midway, occurs between making and remaking and is the methodological and practical linkage between iterations of camp or midway. Midway spaces, like camp spaces, accommodate both arrivals and departures. And as camptown, Gibsonton makes room for the spatial conventions of the midway in its yards. Within these symbolic constructions, building details demonstrate the paradoxical coincidences of the temporary and the permanent as well as the itinerant hesitation typical of Gibsonton's carnival performers. In one case, blocking supports the trailer hitch of a mobile home that has not been moved in forty years. This constructional method is also conventionally found in temporary midway foundations and in single supports and matrices of smaller supports for roller-coaster rides. In its abstraction as a "field condition," the latter system of support corresponds to that "great engine of discovery" that allows us to understand a logic behind the paradoxes found in the town of Gibsonton.[56] As a nonhierarchical, even paratactic, built environment, the camptown relies on interactions among its array of performative spaces for its consistency as a community—a pulling together that allows for continuous arrival and departure.

The time of camping is one of duration, such that "breaking" camp is woven together with "siting" camp through the reciprocal acts of unpacking and packing. It follows that the procedure of camping is fluid and thus neither temporary nor permanent. In Gibsonton this idea of parasitic time as duration is qualified by the seasonal work of the carnival performers and by the community's semipermanent structure. Moreover, the interrelationship between camp and midway conceptually generates a movement-in-place.[57]

"Breaking" camp in Gibsonton in effect means resiting home along the midway. This perpetual process of lingering and awaiting departure constructs a time to which linear, cyclic, or calendric models cannot be exclusively applied. Traditionally, performers and sideshow operators depart in early May for the carnival season; but evidence of a strict adherence to calendar days is found only in a derisive term describing carnival neophytes, eager to sign on, as "Firsts of May."[58] The time of Gibsonton is the intercalary lived time of experience between arrival and departure. And Gibsonton's timing, as camp, requires the redefinition and recalibration of "temporary," "permanent," and "impermanent" through duration and mobility rather than moments or snapshot positions. Resonating with the Bergsonian "undivided mobility of the real," camp conceptually begins as camping, a procedure of fluid place-making that is "*of* and *in*" time.[59] Camping thus precedes the necessity of the campsite, which as we have seen does maintain qualities of movement and duration by way of its parasitic mode. Stability is exchanged for mobility and continuous for discontinuous.[60] Opposing the process of substitution that Henri Bergson's fluid model also seeks to overcome, the parasitic situation of camp facilitates these "changes of state."

The spaces made within Gibsonton's yards are not abstractions, but instead are based on things, kinds of things, and the relation of things. Such spaces and their symbolic potential do not rely on exteriority, but occupy the middle ground, literally the "midway" itself, between the conditions of itinerancy. What Bergson refers to as duration, in Gibsonton's case, is a kind of home, which relies on direct experiences in time and space to signify its organic meaning.[61] This situation yields a condition that is "normal" in places like Gibsonton. Because of the ever-present, essentially quotidian, presence of the midway event, we do not lose the sense of the event itself and thus do not resort to substantive explanations that might fix the ground and its symbolic potential. Concurrently, the yard itself remains open to the exposition and visual experience of kinds of things, such as the momentarily "grounded" carnival equipment.

In Gibsonton, social life is not fixed within what Bergson calls "a projection in homogeneous space." Combined with heterogeneous yard spaces, the fleeting nature of the midway remains in the "perpetual state of becoming."[62] The apparatus of the midway, as a reconstructed performative space, although actually private property, is experienced as public space and serves as a medium for social interaction—a phenomenally defined contract in which the "living self" does not become a "fixed form" of linguistic production.[63] Gibsonton's lived symbols continue to allow for "free action"—a process of making home in which speculation about an indefinite future relies as much on its flexibility as its codified rigors—in Gibsonton's case, seasonal work, zoning ordinances, and other local exigencies. Bergson summarizes the qualities of the town's spaces: "This reality is mobility. There do not exist things made, but only things in the making, not states that remain fixed, but only states in the process of change."[64] In Gibsonton, the symbol is "mobilized," and the "mobile home" arises from a conjectural foundation raised on the paradoxical experiences of yard and midway.[65] And the yard, as homesite, becomes the "lived symbol," at once ambiguous, legible, and livable.

Chapter 9

SLAB CITY

Heterotopic Zones of Domestic Exile, Homelessness, and Encampment

Siting Camp

> The heterotopia is capable of juxtaposing in a single real place several spaces, several sites that are in themselves incompatible.
>
> —MICHEL FOUCAULT (1986)

Slab City's recreational vehicles float on the desert sea. Anchored by its eponymous platforms, the settlement combines military camp, holiday camp, and homeless camp. In a geography where canals become viaducts, siting is a complex dialogue of orientation and navigation. If the site for Gibsonton's camp is the reinvented carnival midway, then "slabbers" find home within the gridded rigors of military administration, barracks, and parade grounds. And yet, ironically, Slab City's martial traces provide a self-regulated, unfettered ground—a result not of zoning ordinances as in Gibsonton, but of marginal, even strategic, siting.

Slab City is located in Imperial County, California, between the Coachella Canal on the east and the Salton Trough on the west. Although for the most part hidden from view, the dominant linear element that defines the area on the west is the southern sector of the San Andreas Fault zone, which branches into the Brawley, Rico, and Imperial faults. More visible, though at a distance, is the Chocolate Mountain range to the east. In contrast to the three-thousand-foot heights of the Chocolate Mountains, the site for Slab City has an elevation of eighty-two feet at its position on East

Mesa. At this close range, the site is defined to the west by the East Highline Canal, the ancient shoreline of Lake Cahuilla, with a shift of elevation from sixty feet to sea level, and the high-tension power lines running parallel to the canal. At the immediate eastern edge of the Slab City settlement are the Coachella Canal and the administrative boundary of the Chocolate Mountain Aerial Gunnery Range, a U.S. Naval installment.

In its historic usage, the Slab City site has undergone at least three phases of organized occupation. From 1000 to 1500 AD, the Salton Bluff Indians sited their camps along the shoreline of Lake Cahuilla. This nonextant freshwater lake included the basin that is filled by the Salton Sea today. The ancient beach line is evident along the escarpment forming the geographic transition between the Coachella and Imperial valleys' sea-level elevations to the West and East Mesa, on which Camp Dunlap and present-day Slab City are sited. The siting of Indian camps along Lake Cahuilla was the result of this geographic feature and the area's abundance of natural resources including fish, clams, and water. The second phase of siting camps in this area was the installation of Camp Dunlap in 1942, formed as a naval training base for World War II. Slab City, the third iteration, is sited on the remaining concrete slabs of the decommissioned military base.

Re-siting Camp: Military Fieldwork and Encampments

The siting and construction of Camp Dunlap was the first human-made intervention into this landscape since the American Indian camps in mid-millennium.[1] The War Powers Act of 1941 allowed the military to move quickly in appropriating the site for the camp, and a Declaration of Taking was filed in California's district court system on February 6, 1942.[2] The installation was activated on October 15, 1942, by the Marine Corps, and its structures were dismantled over time until the Department of Defense finally conveyed the parcel to the State of California in October 1961. The contemporaneous purchase and development of Rancho Margarita, which would later become Camp Pendleton, serving as the primary headquarters of the marine divisions on the west coast, reduced the scale, permanence, and overall importance of Camp Dunlap.[3] Military leaders chose the site for its remoteness, which allowed high-angle gunfire to a distance of up to six miles. Topographically, the site offered a mesa of high ground between two washes, which were later incorporated into Siphons 7 and 8 to help divert water flow from the Chocolate Mountain watershed over the Coachella Canal and around the main campsite. Electricity came from the Imperial Irrigation District's power grid, and freshwater was supplied from the East Highline Canal. Water from the canal was filtered and then stored in two one-million-gallon tanks built within the overall wire perimeter of the camp.[4] A complete sewage-disposal system was installed within the camp, and sixty-seven manholes remain throughout the site.[5]

Camp Dunlap, along with other military camps, was located in this desert context to simulate the conditions of the North African desert in the training of the Marine Corps units.[6] To replicate this situation, military planners modeled the installation on the tent camp layout for military camp and bivouac areas. The layout plan included permanent building types to house the administration, the officer of the day, and the dispensary. Additional permanent buildings were a series of mess halls with their attendant storehouses, toilet buildings, and lavatory and shower structures. For recreation, the main camp area included a projection booth and stage and an anomalous pool interrupting the layout's hierarchy at the southwest edge of the central parade ground. All of the permanent buildings were constructed on the concrete slabs that still remain on the site. Military building contractors maintained a special field laboratory in the camp during the construction of the slabs to monitor the concrete mixes and to check the proportions of the locally available aggregate. This laboratory was necessitated by the extreme heat of the environment during the summer months of construction and by the combination of very fine alluvial clay and sand classified geologically as "Superstition Gravelly Sand."[7] When the camp was decommissioned in 1946, military officials permitted the relocation of the more solidly constructed buildings to Niland, where they have remained.

In terms of these components and their layout, the diagrammatic layout of the tent camp published by the U.S. War Department in 1943 can be compared to the actual configuration of Camp Dunlap shown in the map dated June 30, 1943.[8] Because of its use as a training facility, the camp had fewer officers' quarters than typical field camps. At Camp Dunlap, this officers' area was composed of an administration building and a structure for the officer of the day. These buildings still occurred at the "head" of the layout as in the tent camp diagram, but they were shifted from their typical location along the central axis of the "parade grounds" so that they were aligned with one of the mess hall–storehouse groupings. This alteration subverts the internal hierarchy of the camp and did so as a result of the swimming pool's position along the central axis of the parade grounds. In fact, the Olympic-size pool established the main axis for the entire campground such that the company streets were roughly symmetrical around this line. The symmetry of the overall camp was broken by the depression of a wash cutting into the mesa. Otherwise, the diagram of the military's standardized tent layout was followed. The company streets were formed by paired drives with blocks of centrally located latrines and lavatories on either side and with alternating blocks of mess halls and storehouses. Selected blocks were left open to allow for the pitching of tents. The central grounds were used for the pyramidal tents that housed eight individuals per structure. While proportions among the component areas remained consistent between the diagram and the camp layout, the dimensions of the central parade grounds have been increased by a factor of two to accommodate any necessary expansion of the camp's military training mission.

Clearing Camp

Slab City's location places it in the context of what was historically considered the United States frontier. The meaning of the Spanish word *la frontera*, which denotes both border and frontier, summarizes the nature of this liminal space. Historically, the U.S. frontier has been both line and zone. Initially designated as the "Fall Line" to inhibit uncontrolled growth, this frontier boundary mandated by George III in 1763 was soon superseded by continually expanding growth rings designated by the temporary and sparsely scattered military forts of a newly independent nation. For the most part ineffective, these forts oversaw the unsurveyed and overlapping territories of Spain and the United States. As a result, a buffer zone rather than the precision and unmistakability of a line separated the settled zone from the uncharted territory. The outer limit of this territory was the declared, though still ambiguous, frontier line. The expansive land of the Louisiana Purchase (1803) might be interpreted as the initial and largest buffer zone between settled and unsettled territory.

The buffering and use of intermediate zones has had a long history in the division of lands. Cases in which no natural boundary exists have required what Lord Curzon, in his 1892 analysis of frontiers, called the "Neutral Zone to define Artificial Frontiers." These intermediate zones have often later been formalized into autonomous states, as in the case of the defensive "marks" of medieval times dividing territory and safeguarding frontiers between territories of Charlemagne and Otto. In North America, with subsequent treaties and purchases, the space of discrepancy between unsettled country and the declared frontier line of mandates began to shrink. This contraction of the frontier began with the annexation of the Republic of Texas in 1845—leading to U.S. negotiations with Mexico on borders and buffer zones between the Nueces River and the Rio Grande and subsequently between the Rio Grande and a line three miles to the interior. With the advancements in the technology of surveying equipment, the borders and for that matter the United States frontier continued to pull back and in effect to close.[9] However, markers used to identify the border continued to make the border *line* more like a frontier *zone*, the mobility of the border monuments exacerbating this tendency. The monuments consisted mostly of piles of rock and could be relocated by landowners eager to extend their holdings. Partially as a result of the continued ambiguity of the line, on June 25, 1897, President William McKinley announced the creation of the International Strip, a sixty-foot-wide buffer between the United States and Mexico along International Street in Nogales.[10]

With America's geographic frontier long since effectively closed, only personal frontiers remain for those camping at Slab City. In addition to providing a continuation of the city's community online, the primary stated purpose of its official Web site is the "preservation of Slab City." Since the beginning of 2003, the directors and managers of the site have been preparing to file for classification as a 501(c)(3) organization. Eric Amptmeyer, the Webmaster and primary advocate for the city, notes in a bulletin board posting:

> We need to preserve what might be the last frontier left in America, Slab City. Such a small tract of land, yet the historical importance is quite rich. . . . Not many places are left in this country where a person can just pick a spot and park indefinitely. . . . This is one the last few undisturbed places left in the United States where a person can feel totally free.[11]

In addition to this utopic realization of the American dream in the frontier's traces, a more subjective frontier also delimits and regulates space for the camping public, loosely codifying interaction and the establishment of boundaries "on the ground." W. H. Auden describes this personal frontier in the postscript to his poem "Prologue: The Birth of Architecture":

> Some thirty inches from my nose
> The frontier of my Person goes,
> And all the untilled air between
> Is private *pagus* or demesne.
> Stranger, unless with bedroom eyes
> I beckon you to fraternize,
> Beware of rudely crossing it:
> I have no gun, but I can spit.[12]

If the frontier is ultimately reduced to this human scale of detail, then campsites themselves, particularly within the fluctuation of Slab City's population, become zones of contention between known and unknown, conditions of home and exile, and place and placelessness. Each iteration of camping, as well as each moment of interaction, becomes important in understanding the forces and relationships involved; and it is within the context of the frontier that these "moments" might be understood at Slab City. The frontier, like the campsite and the activity of camping itself, serves as a "mediating ground"—creating a field for performance and interaction and an interlocutory process working between occupant and place.[13]

Debates about the history and evolution of settlements in the frontier of the American West illuminate Slab City's contemporary place through other self-regulated occupations of the internalized frontier of open public land. The argument that the western frontier had its origins in a network of "planned communities" counters Frederick Jackson Turner's popular and long-held idea that the frontier progressed through the gradual concretization of temporary occupation into permanent communities.[14] Outlined in his address at the World's Columbian Exposition of 1893, Turner's thesis was that the westward expansion moved from the hunter-trapper sheltered by a crudely wrought cabin, to the farmer, to the entrepreneur who transformed village to city.[15] But the establishment of urban settlements actually stimulated, rather than followed, the opening of the West to agriculture: "as vanguards of settlement, towns led the way and shaped the structure of society rather than merely responding to the needs of an established agrarian population for markets and points of distribution."[16]

These communities, which included Spanish *pueblos*, mining camps, and railroad communities, were for the most part highly planned and included extensive fieldwork. The general procedure for the settlements began with the selection of a "promising site," the surveying of streets, lots, and open spaces, and the erecting of buildings in predetermined locations. In the particular case of the mining camps that historian John Reps has included in this type of planned community, an overlay of parameters and regulations followed an initial improvisation. Early mining camps "did develop more or less spontaneously, although clearly not as the logical result of earlier pioneer or farming settlement." In spite of the lack of structure, "efforts were quickly made to bring some necessary degree of order through surveys of streets, adjustments in property claims, and reservations" for public buildings and spaces.[17] In its evolution from an improvised settlement to a self-regulated community, Slab City follows this argument for the planned community as the catalyst for operations, in this case for tourism and squatting rights, within the "frontier." The place of Slab City stimulates both dwelling and tourism in the form of Canadian vacationers and retirees who return each season as "snowbirds."

The typology of the military fort offers architectural expression to the paradox of enforcement and open expansiveness along the United States' frontier, in its historical or contemporary permutations. The United States built military forts in the first half of the nineteenth century in an attempt to mark the frontier boundary. As a result of the frontier line's rapid progression across the country, the majority of permanent forts were built in the American West in the 1860s and 1870s, when most of the resistance from Indians and Mexicans was encountered as the frontier reached its final location in the buffer zone of the international border. The military forts were used mostly for the Mexican-American War, after which they often functioned as trading posts. In this dual use, the military forts were the first centers of community and exchange in the southwestern region. On the other side of the border, the Mexican forts, with their "presidios" as the plaza garrison, provided structure and protection for the community's central open space on the frontier.[18] The design of early forts on the United States' frontier emphasized the perimeter wall because protection was a necessity against constant Indian attack. Later, with constructions such as New Mexico's Fort Union in 1866, the buildings were organized around a central parade ground, and the wall was articulated by the arrangement of the officers' quarters, soldiers' barracks, and offices around the edge of this open space. At Fort Sumner, built in 1862, the original wall took the form of an exterior covered walkway along the length of the barracks. This walk, along with a back wall, formed an envelope of space in which the barracks, courtyards, and offices could be arranged.

The double enceinte walls of early French fortifications clearly articulate this envelope of space critical to the understanding of Slab City's military and social boundaries. At Carcassonne, the city walls were refortified in 436 AD in the tradition of Roman building and were again altered in 1285 with the medieval additions. At Montagnana, in the 1360s, a pomerium was added as a streetlike open area running parallel with and adjacent to the interior base of the walls.[19] This "no-man's-land," as wide as a town's street, was cleared between the two walls to keep "the interior surfaces of the heavily buttressed walls clear of obstructions and [to facilitate] the arming of the open-backed towers."[20] In the U.S. military forts, the medieval fortification walls were morphologically transformed by elements moved into the actual wall envelope. The outer, more extensively fortified, barricade included elements such as towers and buttresses attached to its outer edge. Towers also protected and supported the corner bastions of the enceinte. The ground in between offered access to these towers and could serve as a second battle zone if the first fortifications fell.[21]

The site of Slab City can be read as a contemporary transformation of the early American military fort. Its siting over the remains of a military training camp gives additional meaning to this connection between military and retiree camps. The camp at Slab City recasts the frontier fort by internalizing and embedding its fortifications such that the trailers and recreational vehicles become individually defensible vehicles dispersed throughout the grounds. The strategic deployment of these privatized vessels is based on the complex social relationships within the camp and the layout of the remaining slabs. These movable placements become the operators for an "art of the weak." This term "art of the weak" deftly contrasts with the much-used "art of war."[22] And Michel de Certeau differentiates the weak tactic from war-powered strategy, which is "the calculation or manipulation of power relationships that becomes possible as soon as a subject with will and power can be isolated."[23] If strategy relies on spatial manipulation and domination, then tactic looks to nondelimited places for the siting of its practice.[24] In Slab City, the trailers, unable to dominate the space of the camp, must rely on their tactical mobility to create a suitable living space. This space can arise only out of suitable placement, with place as a type of location, and configuration of places, where place results from a multivalent complex of forces, which may be social, environmental, or pragmatic conditions. Tactics allow for an autonomy and methodology that does not rely on the delimitation of an external territory, such that tactics must occur within the "space of the other."[25] In military operations, strategies are far-reaching overlays that occur outside of the enemy's field of vision, while tactics occur within this field of vision and must afford an "invisibility" through their speed, mobility, and "weakness."[26] At Slab City, the compulsory terrain is the manipulated ground of the slabs. In de Certeau's model, we find that tactics "must play on and with a terrain imposed on and organized by the law of a foreign power."[27] Organized law serves as the requisite administrative structure that initially imposed and currently maintains the gridded township system.[28] Historically, military camps supplemented their internal regimentation with a protective barrier of ditches and bastions. Simulating the protective walls of the fort, this manipulation of the ground to create a defensible territory cannot and does not occur at Slab City. Instead, a "weaker" tactic is employed, one that resonates with the shaping of a camp by a ring or circle of carriages.[29] This type of enclosure utilizes the vehicles themselves as a form of protection and thus acts as a precedent to the type of territorial assemblage that is found in Slab City. The "wagon wheel" formations of parked trailers at Airstream rallies simulate the wagon circles, made into a well-known image by early films in the western genre in which the pioneer settlers created a ring to fortify and protect family and possessions. At Slab City, the wheel has been dispersed into more individualized encampments that are influenced by the specificity of the site itself, a particularity that arises from its history, culture, and physical context—that is to say, from the qualities of the place.

The federally mandated division and organization of western lands followed the early settlement of the United States' frontier. These federal actions utilized the grid and systematic tools of partitioning as a planning apparatus. In an unlikely contradiction, this strict organization of lands allowed for Slab City's formation. In the Land Ordinance of 1785, which divided the western lands into a grid of six-by-six-mile townships, a provision set aside section 16 "for the maintenance of public schools" within each township.[30] This provision was reflected in the subsequent Land Act of 1848 that organized the Oregon Territory. In this act, section 36 of each township—in addition to section 16—was to be used for educational purposes. The Treaty of Guadalupe ended the United States' war with Mexico that same year and resulted in California's statehood two years later in 1850. The surveys of southern California that were carried out between 1854 and 1856 incorporated the idea that both sections 16 and 36 should be set aside for educational development or investment. Slab City's future location fell within the boundary of section 36.[31]

More recent land-management policies have also influenced the public use of western lands in ways that reflect the current occupation of the Slab City site. The U.S. Department of the Interior's Bureau of Land Management (BLM) oversees a policy that allows long-term camping in sections of public land designated as Long Term Visitors Areas (LTVA). Established in 1983 to meet the needs of winter visitors, this program administers an array of LTVAs in southeastern California and western Arizona by offering limited security, occasional access to a water supply, and, in a few cases, sewer waste stations. LTVAs differ from typical campgrounds in their lack of services (no electricity or telephone connections), in their expansiveness and dispersal of trailer sites, and in their low cost. Administered by the BLM, LTVAs require the purchase of a permit for a relatively nominal fee. In 2001, a seven-month stay (between September and April) cost $125, and a seven-day period of residence cost $25. Although classified as "long-term," restrictions ensure that terms are limited, and campers must move their campsite to another area at least twenty-five miles away after fourteen days at one location.

The BLM has generated a set of rules for camping activities within LTVAs. The BLM's rules specify a minimum distance of fifteen feet between trailers in adjacent campsites and disallow any "fixed structures," including fences, dog runs, windbreaks, and storage units. Commercial activity within the LTVA requires a vending permit, and quiet hours are between 10 p.m.

and 6 a.m. In addition, campsites cannot be left unattended for a period of more than five days. The Bureau of Land Management does, however, allow for campers to stay for free on federal land outside of designated LTVA sections, but the fourteen-day relocation rule still applies. This program was formulated in 1983, when places like Slab City were becoming popular and the increasing influx of winter campers, particularly between September 15 and April 15, required a degree of regulation by the government.[32]

Slab City is outside of LTVA jurisdiction but shares cost-free camping and the lack of restrictions governing campsite location. The main difference between the LTVA program and Slab City lies in the City's background history, the self-regulating complexity of the society that has evolved there, and the mixing of permanent campers along with the seasonal tourists.[33] But the residents of Slab City, known as "slabbers," do follow the LTVA practice of "dry camping," popularly known as "boondocking." Participating in a phenomenon that now includes sites such as Wal-Mart parking lots and empty urban lands, boondockers camp independently of conventional campground infrastructure, sometimes without water and within a more geographically dispersed social network.

Making Camp

The making of the camp at Slab City differs between its temporary and permanent residents. The retirees and vacationers who occupy the camp from October to the first of May utilize the mobile infrastructure of their camping vehicles. The interstitial space between recreational vehicles and trailers is draped with fabric for shade and privacy and is organized by the placement of furniture and cooking equipment. In some cases, American flags, Astroturf, and camouflage netting enhance these spaces between vehicles. For the more permanent, year-round occupants, immobile or immobilized vehicles serve as the basis for constructing shelter. Both temporary and permanent campers share a need to improvise bricolaged dwellings: "the permanent residents of Slab City seem to be collectors and builders. They start, typically, with a broken-down yellow school bus and begin adding lean-tos and rooms. Next comes a fence or old tires to mark their domain."[34]

As a result of the site's distance from resources, many of those camping at Slab City are gleaners. Both permanent and temporary residents of the city participate in this practice of collecting that results from necessity and is sometimes engaged in as a leisure activity. Gleaning includes gathering vegetables left behind from mechanical harvesters that tend the fields in nearby Imperial and Coachella valleys. Gleaning is also carried out closer to the slabs in the naval bombing range across the canal. In this location, campers collect military remnants for their resale value as scrap metal and recyclables. Copper shell casings and exploded metal fragments can be found in the bombing range—materials that have been incorporated into the city's living structures. In one case, a five-foot bombshell casing has been used as one of the supports for a provisional porch added to the side of a mobile home.[35]

Communication and communal relations are made through the apparatus of the citizens band radio, with each channel providing access to necessary services. Channel 5 is used to place an order for water; Channel 3 allows communication with Don Smith, an advocate for the "bush-bunnies," a disenfranchised group living south of the slabs; and Channel 23 is reserved for ordering propane, calling a mechanic, or listening to the local news at 6 p.m. In the 1980s and early 1990s, the news was presented by Sheila Cox; today Linda Barnett is responsible for the news service. Slab City's Web site also reflects the services available to the community—library (Rosalie at "Drop 7"), Christian Community Center (Rev. Phil Hyatt), Avon lady ("Lady Sundance," located in "Far Area 2"), auto repair (Tommy Flintstone), soft drinks (Area 3), auto electrical service (Nutty Brothers, next to "Builder Bill," in Area 4), and an underground shower at the first guardhouse. This network of trades and services, available both on the Internet and within Slab City's built environment, frames a semipermanent community linked through the technology of the CB radio, and now through the apparatus of the Internet.

Making Camp: From Temporary to Permanent Autonomous Zones

> Burning Man is . . . something like a physical version of the Internet.
> —BRUCE STERLING (1997)

Known informally as "Burning Man," the Black Rocks Arts Festival is held annually in the remote desert of northern Nevada. Larry Harvey, the founder of Burning Man, identifies the key elements of the festival as the experience of labor and the experience of play.[36] Framed as a carnival of the arts, these two components combine in the collective making of the participatory exercise of Burning Man and the construction of the spectacle and sculpture of the "Man" itself. The festival concludes with the conflagration of the Man—the repetition of an event initiated by Harvey on a California beach in 1986. Members of the Cacophonist's Society helped Harvey develop Burning Man's present form in 1990 with what they called "Zone Trip #4." The organizers of the Zone Trip and Burning Man understood culture as something that was *made* through participation and experience. Harvey notes that the transformative experience of the event represents "not art about society but art that generates society." Framing itself as a noncommercial event, Burning Man has allowed the sale of only two items: ice and coffee. In 2002, according to the Bureau of Land Management, an estimated twenty-nine thousand participants attended Burning Man.[37]

Like Slab City, Burning Man is sited in a desert area. Both locations share a similar climate and degree of remoteness. Like Slab City's current and historical use as a naval training and bombing region, the Black Rock Desert was used in the 1940s and 1950s as a bombing range and currently functions as the site for low-altitude aviation training runs. Both zones are also characterized by basins of internal drainage—the Salton Sea was formed as the water from the breach of the Colorado River canals traveled to the area's lowest point, and the Black Rock zone collects drainage from surrounding highlands. But geographically, Slab City is at the edge of the Salton Trough and the natural features of the ancient beachline and the line of Chocolate Mountain range. Burning Man's site is located in the middle of a basin and characterized by the emptiness and flatness of Lake Lahontan in the Black Rock Desert. Recharged with water in the winter months, the lake bed dries during the summer, forming a flattened silt alkaline slab of cracked earth. The winter moisture flattens the surface sediment into hardpan and erases all traces of human occupation. The playa, the focal space of Burning Man, takes its name from this lake and the geological characteristic of the silt alkaline lake bed. Like the Long Term Visitor Areas (LTVA), this area is administered by the Bureau of Land Management.

The Black Rock Desert is also the site of land-speed records and the natural phenomenon of transient dunes. On October 15, 1997, the British

Thrust SSC Project set the first supersonic land-speed record with an average speed of 763.035 mph. The nearby Bonneville Salt Flats in western Utah hosted the "World's Fastest Mobile Home," documented by photographer Richard Misrach in 1992.[38] Recently, the Winnemucca Field Office of the Bureau of Land Management issued a travel advisory as a result of a series of unexplained transient dunes during the summer months of 2002.[39] The BLM subsequently mapped and documented the movement of the dunes, which amplify the impermanent ground of the camping experience.

To form Black Rock City, the Burning Man festival is laid out as a two-thirds circle, with nine concentric semicircles on its interior. Clock time and degree-based coordinates are used to locate theme camps and other features within the city. Main axes occur on each "half hour" between 2:00 and 10:00 (60° to 300°). Center Camp includes the main "playa" area, and the sculpture of the Burning Man is located one-quarter mile north of Center Camp, also called Camp Headquarters, at the circle's geographic center.[40]

The theme camps of Burning Man identify the social sectors and overall organization of the festival. These thematic groups are similar to Slab City's makeshift cliques of Canadian snowbirds and its singles' club called Loners on Wheels (LOW). Historically, the Midway Plaisance at the World's Columbian Exposition of 1893 included a series of thematic camps such as the military and Bedouin encampments as well as Sitting Bull's camp. The Exposition was also comprised of a logger's camp, a hunter's camp, and an Australian squatter's hut in the main grounds of the fair. Just as navigation of the Midway Plaisance and the Exposition required a series of guides, each year the organizers of Burning Man publish a set of criteria on the official Web site of the event, which also includes a "Theme Camp and Village Resource Guide" that provides information about city layout, packing supplies, and "unregulated self-reliance."[41] Theme camps from 2003 included Carn-Evil Camp (sideshow, carnival midway, and fun house), Christmas Camp, Free Photography Zone, Bad Idea Theater, BRC Bike Repair and Divinity School, and Lost Temple of Waterboy. The following section from Burning Man's Web site identifies the locations available for the theme camps and gives an idea of how the overall camp is organized:

> The Esplanade—This is the first street at the front of the city and

faces the Man. It is reserved for camps that have 24-hour interactivity, a completely conceived visual scheme, and playa-frontage needs. (Evaluations based on past project history, adherence to deadlines and clear visual plans.)

Non-Esplanade—Located within the city limits. This is for fun interactive camps without need for playa frontage. High visibility on the streets behind the Esplanade.

Center Camp—Very highly interactive camps that fit into the center camp "groove" with 20 people or less.

Large Scale Sound Art—At the ends of the city, 10:00 and 2:00. Reserved for large sound systems which are filled on a first-come, first-served basis.

The site plan of Burning Man echoes the traditional organization of the camp around the campfire. The site plan is a radial plan with the Burning Man—as sculpture and the event of its burning—at its central focus. In a diagrammatic reading of the campsite, this arrangement reflects the idea that camps, distinct from military camps and fortifications, are not enclosed but instead radiate out from a center like the light of the fire. Burning Man's circular form and its theatrical environment hold religious significance in their circumscribed definition of American sacred space: "sacred meaning and significance, holy awe and desire, can coalesce in any place that becomes, even if only temporarily, a site for intensive interpretation."[42] This intensity of sites relates directly to discussions of place as event and also alludes to Burning Man's similarity to Chautauquan gatherings and evangelical camp meetings. Like the participants in camp meetings and spiritualist events, the festivalgoers at Burning Man seek to disentangle themselves from "worldly care" and to experience "sublime truths of revelation" through an escape to the "wilderness" where senses are heightened in the "healthful resort" of the campground.[43] The Burning Man festival is also transformed into a sacred space through the development of what has been called place-myths, "the composites of rumors, images, and experiences that make particular places fascinating."[44] The place-myths of Burning Man are what remain after the idea of "Leave No Trace" has been applied to the site itself. This mandate, authored by the Bureau of Land Management, asks that campers on public federal lands in general and Burning Man festivalgoers in particular remove all materials and evidence of their occupation to preserve the nation's cultural and natural resources.[45]

Breaking Camp: Heterotopic Camping and the World Wide Web

> I've been monitoring Slab City on the net, as it's the only way I HAD for visiting. But now I'm retired at 52 and understand it's the state that's going to shut it down?
>
> —DANIEL ADKINS, post to slabcity.org (December 28, 2002)

Slab City as slabcity.org and Burning Man as burningman.com share a presence on the Internet that has become typical of camping culture. Burning Man's festival continues through its Web site with discussions and postings about recent festivals and upcoming events.[46] At least three different discussion groups and bulletin boards link Slab City's community.[47] The Internet has thus become the "permanent site" for these communities, and each campsite's HTML code is an integral part of its architecture. Campers at Slab City move from place to place throughout the year, and festivalgoers spend fifty-one weeks away from the grounds of Burning Man. Larry Harvey, the founder of the festival, notes its parallels with the Internet: "Burning Man, then, is a compelling physical analog for cyberspace, and, unsurprisingly, we have attracted many people who regard the experience as the equivalent of cyber-based reality."[48] At the same time, the event of Burning Man is a very real experience, and festivalgoers must prepare for the extreme heat and lack of services at the remote site.

Slab City shares this liminality and the extreme climatic conditions. But in terms of physical presence, the City is a Burning Man event that has duration—an undissolved Black Rock Arts Festival. This combination of presence and absence with the "real" is made possible and is in many ways reconciled by Slab City's layered grounds and its presence on the Internet.[49] While Slab City is not technically a heterotopia, the campsite is heterotopic in its characteristics that may lead to the formation of a heterotopia. Its potential for being a heterotopia-in-the-making locates Slab City in contemporary discussions of "other spaces." It is the layering of places within the site, rather than spatial juxtaposition, that effectively makes Slab City and its spaces. Place here precedes space. In this important distinction, place occurs *in* time as well as *through* time. Slab City combines "festival time" in the former and museumlike accumulation of time in the latter. In such a slippage, the past and the present coexist, albeit uneasily through the layers of the place, which can be identified historically and archaeologically.[50]

In his post to the Slab City message board, Daniel Adkins refers to the vicarious experience allowed for by the virtual camping space, one that might be realized with retirement. Hakim Bey also describes this conflation of camp and virtual space. His formulation of the Temporary Autonomous Zone (TAZ) as ephemeral "uprising" builds on the possibility that such camps rely on clandestine, sometimes virtual, nomadic routes that can only be truly developed and mapped in cyberspace: "If the TAZ is a nomad camp, then the Web helps provide the epics, songs, genealogies and legends of the tribe; it provides the secret caravan routes and raiding trails which make up the flowlines of tribal economy; it even contains some of the very roads they will follow, some of the very dreams they will experience as signs and portents."[51] The Web presence of these camps on the Internet forms countersites that reflect, in a virtual context, the marginality of their location and their social-cultural construction. As countersites, these camps, in their manifestation on the Internet, are heterotopic constructions that serve as a "bundle, cluster, and set of relations."[52] But the expressions of these camps in the physical world and on the Internet share a construction of the place as event. As a result, virtual and material architectures converge in a tactical practice of making place within the externalized frameworks of the Internet and the land policy that has allowed the formation of Slab City. Camping itself serves as an appropriate analog for navigating and operating within the places constructed on the Internet, each of us operating from our own virtual campsite. Slab City and Burning Man point toward other contemporary concerns of a camping life. The slabs of the desert camp and the platforms of Manila Village anchor experiments of place, but are susceptible to political and environmental winds of change. And the thematization and formalization of camps has implications for urban identity and urban settlement design, particularly after natural disasters. It is ironic but not surprising to find a theme camp called "Camp Katrina" established by Burning Man participants two thousand miles away in southern Mississippi.[53]

Chapter 10

FROM MANILA VILLAGE TO NEW ORLEANS

Asymptotic Territories of the Mississippi Delta

Camp of Ships

> Now when it was not yet dawn, but night was still between light and dark, then about the pyre gathered the chosen army of the Achaeans, and they made about it a single mound, drawing it out from the plain for all alike; and by it they built a wall and a lofty rampart, a defense for their ships and for themselves.
>
> —HOMER, *Iliad*, 7.430–41

This assemblage of the Achaeans' "hollow ships" forms a camp at the mouth of the Maeander River during the ten-year siege of Troy. The Achaean army initiates the formation of the camp with a ritualized burning of the battle dead in a funeral pyre. Around this central feature of the pyre, the Achaeans' camp becomes a temporary city that resonates with the walled and fortified city of Troy. This "city of ships" establishes an order between fixity and movement that it shares with ritualized traditions such as weaving and dancing in ancient Greece.[1] The Achaeans, through the process of burning the dead and arranging their landed ships, have woven and remade the grouping of ships into a camp that is both temporary city and fortress.

In Louisiana, the Delta region of the Mississippi River and its ecological partner the Atchafalaya Swamp include settlements that must account for a landscape that is predominantly water and malleable sedimentation. Like the landed ships forming the Achaean camp, a history of houseboats both fixed and mobile characterizes the inhabitation of the Atchafalaya and the Delta. The decks of the houseboats served as a reconstituted ground on which chickens were raised and gardens were cultivated.[2] In particular, the frequent flooding of Bayou Chene in the Atchafalaya Swamp forced many of the area's residents to relocate onto houseboats.[3] This "landing" of a house made temporary by natural forces of flooding and by a landscape whose principal feature is water has occurred in the Mississippi Delta settlements of Manila Village and, more recently, in the postdisaster urban environment of New Orleans.

Siting Camp: Water

> The earth floats on water, which is in some way the source of all things.
>
> —THALES OF MILETUS

Water serves as origin of and support for a floating terrestrial plate, a platform. In the Old French, the word *plate-forme* denotes a diagram. Itself an epistemological *plate-forme*, the aphoristic statement of the pre-Socratic philosopher Thales diagrams two interactions of "earth" and water. The first association, which is explicit, describes the vertical and sectional relation of the earth to water. Thales believed that the earth, composed of matter with a nature similar to that of wood, floated on water. In this way, earth is a platform resting on the water's surface. The second relation, which is implicit, is a horizontally oriented diagram of the origin of earthly matter in water. Living in peninsular Miletus, where he had founded his school of natural philosophy, Thales was in the position to observe the generation of earth from water as the Maeander River emptied its sediment-rich flow into the Gulf of Lade. Across the gulf, the coastal storage houses of the town of Priene had to be rebuilt closer to the receding shoreline during the time of Thales.[4]

Thus, the transition between earth and water at the coastline submits to the fluidity and changeability of water. This sedimentation has resulted in the current location of the town of Miletus almost six miles inland. Thales also explained the phenomenon of earthquakes as the result of the fluctuating dynamic of "rough seas," not unlike the Homeric attribution of natural forces to Poseidon. In the Mississippi Delta region, the phenomenon of *tremblancs* includes the shaking earth of the coastal prairies.[5] The shifting of earth by hydrologic rather than seismic forces defines an exceptional case of grounding without ground in which the sea's horizontally expansive flow and the tide's vertical ebb form an uneasy landscape. These two diagrams, inherent in Thales' thought, provide an initial foundation for the study of the platforms of Barataria Bay's Manila Village in the Mississippi River's Delta region. As semipermanent constructs within the coastal landscape, these platform communities constitute a type of camp, initially understood through the vernacular architecture of the cultural groups who founded Manila Village.

For the nomadic Badjao groups, the prevalence of water in their archipelago environment of the southern Philippines necessitated the development of wooden "boathouses" and "land-houses" that could float on or hover above the shallow coastal areas of the Sulu Sea. Originally relying exclusively on the mobility of the forty-foot-long boathouse, the Badjao subsequently erected supplemental dwellings on stilts that were often grouped together to form artificial islands with networks of wooden planks connecting spaces.[6] Similarly, the Mindanao, known as "people of the floodplain," in the basin of the Pulangi River, and the Tausug, the "people of the current," built groupings of wooden stilt platforms above the shifting tidal waters throughout the Moro Gulf. Construction and layout of the posts for these housing platforms followed a ritualized sequence. The typical Tausug house indexed parts of the body in its construction sequence. Nine poles were arranged in a grid, and the first post set at the center symbolized the navel. Subsequent placement of the supports continued with the southeast corner, representing the hip, and the northwest corner, symbolizing the shoulder. After all four corners were placed, the infill supports followed as rib, neck, and groin. The Tausug builders traditionally practiced this ordering to "ensure the durability of the house and the protection of its occupants."[7]

The platforms of Manila Village combined the connectivity of the floating Badjao shelters with the Tausug people's use of pilings for fixed structural support. Although no longer the floating boatlike units of the

Fonville Winans, Aerial view of main shrimp-drying platforms of Manila Village, Barataria Bay, Louisiana, ca. 1938. Fonville Winans Photographs, Louisiana and Lower Mississippi Valley Collections, LSU Libraries, Baton Rouge, La.

Badjao, the Manila Village structures retained a connection to the shifting level of tidal waters in the horizontal planes' close proximity to the water surface. Thus, one connection with camping that can be made is the "fixing" of a virtually mobile unit, whether considered to be ship-as-platform or boat-as-house. The agglomeration and connectivity of these units occurred through other semifixed constructs such as the walkways, some of which floated while others were raised on stilts. New platform-vessels could easily be added by way of new plank walkways. The water served as a ground for temporary support, which was then replaced by stilts and pilings. The water also provided the sole means of access to and from this community. In this context, water acted as connector and isolator, allowing for and at the same time limiting contact with the mainland. Mobility in effect allowed for the development of a fixed and semipermanent community at a distance from territorial governance.

Clearing Camp: Creolization and Difference

As early as the mid-eighteenth century, Philippine immigrants and exiles, many having escaped the brutality of Spanish galleons, began settling the bayous of the Delta region. By the 1870s, the array of platforms known as

Manila Village had begun to rise above Barataria Bay's tidal waters within the marginalized zone of the sediment-laden wetlands of the Delta region.[8] This zone also attracted Chinese immigrants, who added their knowledge of shrimp-drying techniques to the developing community of the village. During this time, other exiled groups including the "Acadians," or Cajuns, from Nova Scotia, the French Creoles from Haiti, and Croatians from the Dalmatian coast settled this area because of its marginality and natural resources. For the most part, these resources were familiar to the groups, although sited in an unfamiliar landscape. Croatians contributed their techniques of cultivating and "tonging" for oysters—adapted from the rocky Dalmatian coast and the Mediterranean climate.

The cultural creolization of the region characterizes the important aspect of "difference" within the Delta—a diversity that was maintained within the community, itself sited in a reterritorialized zone of contingency. The area's history of contested waters highlights the territorial nature of this Delta zone. Aggravated by the littoral flux and the discrepancy between high and low tides, the fluctuating definition of territorial waters, navigable waterways, and the marginal sea belt locates the site for Manila Village and other platform communities within a district of agency where legal boundaries are indefinite and semiautonomy shares the space of what is effectively a colonial "possession." A three-volume publication by the National Oceanographic and Atmospheric Administration (NOAA) summarizes the technical and legal aspects of determining maritime boundaries in the United States from both historical and contemporary perspectives.

From these documents, it is clear that the determination of "territorial sea" is a highly litigated problem, involving changes and forces defined in meteorology, astronomy, climate, and geography. The two main problems that characterize the situation in the Delta region are questions of delimitation and passage, which were the subject of a United Nations conference in 1958.[9] Because the tidal range varies with the particularity of sites, the problem of delimiting edges of land and thus territorial boundaries is highly dependent on contextual specificity. For example, the Supreme Court's list of factors in defining "a particular island's status" reflects the significance of context: "size, distance from the mainland, depth and utility of intervening waters, shape, and relationship to the configuration or curvature of the coast."[10]

Within this concept of delimitation, two problem zones seek resolution: island and mainland. First, the problem of determining the edge of land through the "ordinary low water mark" from which to begin the territorial sea poses a problem because of the "ambulatory" nature of the coastline. In fact, the Supreme Court included the following characterization in its ruling for the *Louisiana Boundary Case* of 1969: "any line drawn by application of the rules of the Convention on the Territorial Sea and the Contiguous Zone would be ambulatory and would vary with the frequent changes in the shoreline."[11] Nautical charts, as static maps, cannot reflect the ongoing natural processes at work in the Mississippi Delta region and, in such extreme cases of fluctuation, describe the situation only at the moment that the reference aerial photograph was taken: "Again we emphasize that we are not suggesting that the chart, as printed, contained errors. Rather the Court recognizes that with erosion and accretion no chart is likely to remain accurate forever."[12] This "ambulatory line" becomes a wider zone of indeterminacy with the problem of differentiating islands from mainland and in the attempts to reference the line's origination and to define "inland waters." The Supreme Court's responses to the mobility of the Louisiana coastline reflect the paradoxes of this situation, one that must be excerpted fully:

> But again Louisiana's geography varies from the norm. The Supreme Court has described the Louisiana coast as "marshy, insubstantial, riddled with canals and other waterways, and in places consists of small clumps of land which are entirely surrounded by water and therefore technically islands." In other words, in at least the delta areas, the mainland *is* islands. If an area of land surrounded by water at high tide (i.e., an island) cannot form the headland of a bay there are no bays on the Louisiana coast. That conclusion seems counterintuitive.
>
> In fact, the Supreme Court has already determined that some of this marsh land should be considered mainland. In *Louisiana v. Mississippi*, . . . the Court said "Mississippi denies that the peninsula of St. Bernard and Louisiana marshes constitute a peninsula in the true sense of the word, but insists that they constitute an archipelago of islands. Certainly there are . . . portions of sea marsh

> which might technically be called islands, . . . but they are not true islands." *Louisiana v. Mississippi*, 202 U.S. 1, 45 (1906). It went on to treat the peninsula as mainland.[13]

Further complicating the differentiation of island and mainland and resonating with the platforms of Barataria Bay is the phenomenon of "mudlumps" in the Mississippi Delta.[14] This problem of an island's potential assimilation to the mainland reflects the tangential and asymptotic nature of the platform structures. Although the platforms were fixed, the changeability of the estuarine landscape meant that at certain points in time, particularly at low tide, the structures were provisionally connected by land formations to the mainland. The following excerpt summarizes this problem of the "mudlumps":

> *Islands v. Mainland.* After the Supreme Court concluded that features that meet the Convention's definition of "island" may nevertheless be assimilated to the mainland . . . the special master was faced with a number of areas in which that question arose. Most common were the "mudlumps." . . . Typically these features appear just seaward of the jetties that form at the mouths of the river's distributaries. These jetties frequently form the sides of indentations into the mainland. The state took the position [opposed to that of the United States] that the more seaward mudlumps, though technically islands, should be assimilated to the mainland and serve as headlands for bay closing lines.[15]

This juridical discourse provides a framework for developing the watery camp of Manila Village. As "travel literature," the litigation narrative suggests that camping on the bay must proceed by bricolage and must occupy a particularly ambulatory spatial frontier.[16] The spatial stories of court documents and legal proceedings orchestrate the assemblage and the displacement of frontiers. These juridical stories operate on places as an everyday "mobile and magisterial tribunal" for delimiting boundaries.[17] For their part, the places also act and operate on the stories, which are not an untransformed overlay onto the place. The anomalous and "counterintuitive" aspects of southern Louisiana's landscape result in a near-mythical treatment by the litigation proceedings presented above. Unique to the Louisiana Delta, the phenomenon of mudlumps affects the writing of the "tidelands litigation." Moreover, the mudlumps reflect the platform constructions in their marginality and mixing of land and sea attributes and forces. The attributes problematic to the litigation proceedings of the mudlumps are actually the qualities beneficial to the inhabitants of Manila Village. The liminality and shifting ground of the Delta region creates a peripheral condition, making assimilation difficult and allowing for a degree of autonomy outside of the feasible limits of political control and yet protected from the vagaries of the unfettered sea.[18]

Making Camp: Detail and Technique

While questions of territory are important to the diagram of a horizontally mandated situation, a return to the vertical relation between water and platform facilitates scalar shifts to the detail, where the construction and composition of the platforms themselves occur. Chung Fat and Quong Son were two platforms within the complex known as Manila Village. Comprised of housing, warehouses, storage facilities, a post office, and markets, these platforms were effectively networks of smaller platforms connected by elevated walkways. The platforms, built from naturally resistant and readily available cypress planks and hand-driven pilings, rose between eight and ten feet above the tidal wetland. Their construction was inherently tied to their use for the drying of shrimp. Chinese migrants from California had introduced techniques for drying shrimp by 1873 and continued to export the majority of their catch back to China.[19] In 1885, the Chinese immigrant Yee Foo received patent number 310–811 for the process of sun-drying shrimp.[20]

This technique was adapted to the conditions of Barataria Bay, remote in its geopolitical relations but nonetheless close to the New Orleans–Mississippi River transportation network and its shipping lanes. Shrimp caught by trawling lugger boats of French origin were initially off-loaded and boiled in cast-iron kettles or copper cauldrons with a solution of coarse salt and water.[21] The shrimp were then loaded onto the surface of the platforms and, with long wooden rakes, were spread into a single layer for drying. To aid the drying process, the platforms were sloped, with valleys at the junction of the planes, for drainage of excess water. Many of these platforms were expansive; the main drying platform at Manila Village was approximately forty thousand square feet in area.[22] Drying the shrimp took three days in the summer and up to ten days in the winter. The dried shrimp were then raked into circles approximately twenty feet in diameter, and the event known as "dancing the shrimp" began: "The men with arms on one another's shoulder formed a circle and the dance commenced. Round and round they went. Tramp, tramp, tramp, their bodies swayed in unison as the men chanted some unfamiliar tune, all the while treading on the piles."[23]

Often accompanied by music, the dancers wore specially made wooden shoes to crack the shrimp shells efficiently and to remove the heads and hulls, which were then raked across a screen to separate shrimp meat from shell. Modified barrels and baskets contained the shrimp for shipping. Oak barrels, typically used as containers for trade in Europe, were adopted in the shrimp industry, with the accepted standard that one barrel was the equivalent of 210 pounds of "head-on" shrimp. Also in use were baskets identified by shrimpers as "chinees."[24] Their forms and techniques imported by Chinese immigrants, these palmetto or oak-slat woven baskets were also known as "half-barrels," indicating that their capacity of 105 pounds of shrimp was set to the barrel standard. The smaller "champagne basket" was also related to this standard, holding 70 pounds and equaling one barrel when three "champagnes" were grouped. The champagne basket received its name from use in packaging champagne imported to New Orleans from France.[25]

Breaking Camp: Untying the Diagram

The diverse spatial practices and techniques of Manila Village complicate the diagrams, in the vertical and horizontal orientation, that were proposed as a preliminary framework. Serving as icons of "intelligible relations" between land and water, the diagrams also take into account the multiplicity of the platform phenomenon.[26] As an explanatory device, the diagram formalizes relationships, particularly the fundamental confluence of sky, land, and water in a geographically liminal place. But the platforms are the scene of a practice—one that is ritually and materially charged. Camping practices similarly rely on this tension between the formal context of a place, as campsite, and the essential formlessness of camp's spatio-temporal process. A camping practice diagrams experience of place. For the architect Peter Eisenman, the diagram is "not only an explanation, as something that comes after, but it also acts as an intermediary in the generation of real space and time."[27] As a diagrammatic practice, camping does not resolve dialectics of mobility and fixity. The self-regulated, multiethnic community exploits this disconnect to make a place for the unencumbered production of dried shrimp.[28] Manila Village, as camping space, becomes a proliferating landscape organic in its tenuous placement just above the tidal ebb and flow. This village indeed becomes a rhizomatic formulation, the limits of which rely not on material resources but on temporal flux itself.[29]

As Giambattista Vico said, "Thales began with too simple a principle: water; perhaps because he had seen gourds grow on water."[30] Although gourds may not naturally grow on water, the economic confluence of dwelling and subsistence, of production and its medium, was a unique feature of Manila Village's ecology. The political matrix, more rhizomatic perhaps than Thales' imagined gourds on the water, also allowed for the village and requires a third diagrammatic schema to understand the relation of identities and geographies.[31] In their historic traces, these platform communities negotiated paradoxes of proximity and disconnection, relatedness and multiculturalism, and the fixed and the transformed. And the more contemporary manifestation of the platform as oil-drilling derrick invokes a parallel relation to place, resource, and economy. Though similarly resulting in a contested ground that can be studied territorially as a network of adjacencies and points of control, the scale of the derrick constructs themselves complicates comparison to the shrimping platforms in spite of a basic formal similarity. The third diagram—that of the asymptotic territory—can further inform the two diagrams, both horizontal and vertical, introduced by way of Thales' early pronouncement.

This diagrammatic overlay does not inscribe oppositions between nature and technology, political entities, or cultural backgrounds. Instead, the asymptotic territory narrates between two situations: the para-sitic and the tangential. At one level, the Manila Village platforms parasite the sea's resources—they rely on the catch of shrimp and the nearby shipping lanes. But at another, more fundamental level, the platform community's very existence is a function of the complicated geopolitical relation between a shifting landmass and makeshift political boundaries. The platforms parasite the relation rather than the particular system or entity. Historically, the village tied itself to the litigating bonds of another people and place: "To play the position or to play the location is to dominate the relation. It is to have a relation only with the relation itself."[32]

The platform community's parasitism must also be understood as prepositional.[33] This condition allows for the combination of paradoxical qualities of temporality and permanence as well as propinquity and distance. The groupings of platforms parasite the relation between oppositional elements rather than the entities themselves, whether political or institutional. Thus, this condition of parasitism occurs from within the system and is exemplified both by the operative practice developed in the creolized community and in the labile architecture that must similarly negotiate the political and natural landscape. Defined by Michel de Certeau as a "calculated action determined by the absence of a proper locus," the tactics practiced by the platform dwellers included building, dwelling, making, and even shrimping.[34] The resulting "metaphorical" or mobile city, resonating with an archipelago condition, is not unlike a "dark sea from which successive institutions emerge, a maritime immensity on which socioeconomic and political structures appear as ephemeral islands."[35] This is a city that is built to specifications but is subsequently altered by a series of uncontrollable events.[36] The platform city's contingencies for such disaster include its flexible design, its redundant structure on pilings, and the inherent mobility of its inhabitants.[37] Urbanization complicates these contingencies, and we will see how New Orleans, mobilized by disaster, becomes a camping urbanism.

The asymptotic territory of Manila Village also renders a tangential situation. Such is the condition of touching without intersecting. The platform community maintained contact with nature, economy, and institutions, while preserving a distance that allowed for multiplicity. Yee Foo's patent of 1885 and the eventual purchase of several small "islands" by Chinese immigrants from the Louisiana Land Office for $1.25 per hectare indicate this partial appropriation of institutionalized structures of ownership and invention.[38] Two elements are evident in this tangency: convergence and divergence. The asymptotic condition of mixing without combining thus occurred within Manila Village and through its external relation with the "mainland." Emerging from the "nautical noise" and its tidal fluctuations, the platform settlement existed as a series of contingencies—local adaptations that in the end meshed with a global situation. Manila Village was always coming into being. Just as the village relied on its amalgamated practice of shrimp drying and the *mattang* maps relied on indirect, tangential routes, camping is a becoming: "It is the chain of genesis. It is not solid. It is never a chain of necessity. Suddenly, it will bifurcate. It goes off on a tangent. It surrenders to the passing signals, the fluctuations of the sea."[39]

This asymptotic architecture has implications for how we understand the breadth of camping practices at the scale of regionalism as well as at the more detailed scale of actual construction. Recent discussions of the problem of region have noted that "difference" based on autonomous, or otherwise clearly recognizable, regions is no longer possible and that such distinctions must be refocused to understand internal differences within a region.[40] Similarly, the postcapitalist situation has been characterized by an "immersion in process."[41] The platforms of Manila Village required both conditions: the immersive ritualized practice of shrimp production and the place-as-region attributes of a simultaneously contiguous and foreign identity. The region becomes a situation derived from "chains of contingency" rather than a dialectical opposition waiting for resolution. The diagram is thus linked to a utopia, or at least a utopic impulse as in Ruskin, Florida.[42] Consequently, it could be said that platform architecture, with its utopic characteristics, occupies the middle ground between heterotopia and an atopia.[43]

A diagrammatic practice does not have to oppose a tectonic practice.[44] As a bricoleur, the troubadour-architect might work from such hybridized diagrams to construct a built landscape in much the same way that the Japanese carpenter as *daiku* works from the scrolls of details and diagrams to compose the dwelling.[45] Without a complete rendering of what the completed project will look like, this carpenter-architect uses the detail-diagrams to carry out "local operations" that result in a finalized composition or grouping. The complete construction comes about from an assemblage of fragments guided by the relations *between* details and *within* the diagram itself. The final construction arises from a bricolage of components, gleaned from the local environmental conditions and taken from the procedural scroll or diagram. A composer of matter and materials, the architect as troubadour then achieves a degree of mobility to "sing," or "dance" as in Manila Village, the construction of a project from place to place. The combination of diagrams—those of matter or "intelligible relations" and that of the diagram as generator—could serve as a potential ground for situating architectural practice within this asymptotic territory.[46] Perhaps the architect-bricoleur can reoccupy these spatial traces[47] of the platform communities and reimagine the resonance of matter and memory, identity and multiplicity.[48]

Breaking after Disaster: New Orleans and the Future of a Camping Urbanism

> [W]e find that confusion increases, just as a city built on the most correct architectural plan may be shaken by the uncontrollable force of nature.
> —WASSILY KANDINSKY, *Concerning the Spiritual in Art*

Forty years after Hurricane Betsy destroyed Manila Village's platform community, Hurricane Katrina devastated New Orleans and other areas along the Gulf coast. Critics have pointed out the historical errors in siting the city of New Orleans. And while it may not have been built to the "most correct architectural plan," New Orleans is a place, its identities undeniable and its spirit indefatigable. After Hurricane Katrina, an array of responses has been generated from an equally diverse set of respondents. Architectural competitions were formulated, designers of New Urbanism were consulted, and, by the start of 2006, many rebuilding projects were initiated. Between the event and these responses, a much-maligned Federal Emergency Management Agency (FEMA) deployed predisaster and postdisaster shelters, and citizens of southern Louisiana and Mississippi have tried to supplement these provisions with their own resources. Many residents have left as "refugees," while others have remained in an internally displaced status or have resisted displacement and maintained their home base with trailers or through rapid rebuilding. The FEMA trailer itself has become a symbol of mismanagement,[49] and design projects have attempted to build what amounts to a "better box" or to appropriate the trailers in a rhetorical modeling of an impossibly elevated city.[50] Others have made significant and important progress in addressing affordable housing design and its availability within Gulf Coast communities.[51]

But what remains undiagnosed is the significance of the urban camping that occurred after the hurricane and is still taking place in many of New Orleans' most devastated areas and in its extended region. What does it mean to camp in the city? Such practices have not been formalized to any great degree since the nascent automobile culture of the 1920s and 1930s, when the municipal camp provided inner-city lodging for tourists and for others seeking work.[52] In these postdisaster situations, the municipal camp has reemerged and has now become what FEMA calls the "temporary group housing site." These federally planned and installed camps contrast other localized camping spaces—many self-governed and often deploying FEMA-issued trailers on private land. These housing responses at the urban periphery, within the hardest-hit districts, and in some cases of urban infill, have redefined, if temporarily, the urban identity and image of the city.[53] After Hurricane Katrina, New Orleans has become the "migrational city" that Wassily Kandinsky imagined and that Michel de Certeau took up as the antidote to the "clear text of the planned city."[54] But the New Orleans experience since August 2005 differs from de Certeau's residually legible urban environment of the 1970s and 1980s.[55] Not only does the camping urbanism relate to its host infrastructure asymptotically, but our own understanding of the housing responses and their problematics have also been conveyed as fleeting fragments through media reports and blog sites. The migrational city cannot be idealized as the structural counterpoint to urban-renewal schemes and late modernist planning. The camping environment, however—its forms, its failures, and its multivalent futures—can be reviewed to rethink future postdisaster response, not through the vehicles of camp (FEMA trailer, Katrina Cottage, and other designs of housing units) but through the urban camping situation itself.

At 8 a.m. on August 28, the Superdome became a "refuge of last resort." Evacuees were asked to prepare for their disaster accommodation as if they were "planning to go camping." At its peak, the Superdome housed twenty thousand disaster victims before Hurricane Katrina's winds tore the arena-tent's weathering membrane. The subsequent relocation, well-documented by television media, continued until September 4, when all of the Superdome's inhabitants had been relocated to the Houston Astrodome. Having been displaced by the disaster and forced into flight, these citizens of New Orleans were now understood as "refugees." This experiment of sporting

Interior view of Reliant Astrodome, Houston, Texas, September 2005. Andrea Booher, FEMA.

arena as urban camp was not new. In fact, the Superdome was an official disaster shelter in New Orleans' evacuation plan, and the site had been used briefly in 1998 during Hurricane Georges. The density of the arena's population internalized and magnified the problems of camping. Necessary connections to the city were made impossible by the Superdome's centripetal design for spectacle and by the flooded urban context. This hermetic situation rendered any asymptotic possibilities of emplacement within the city untenable.

Other new and unexpected forms of the camp included the cruise ship and the convention center. The decks of a Carnival Cruise Lines ship served as a campsite for seven thousand evacuees on September 4, 2005. The ships *Holiday*, *Ecstasy*, and *Sensation* were also contracted to house evacuees. The use of the cruise ship and the Superdome highlight the double displacement that can occur with the heterotopic situations outlined by Michel Foucault.[56] Already dislocated from their homes, disaster victims then became "refugees" in their uneasy status, away from their home-place. Both cruise ship and Superdome became the "floating piece of space." In their unexpected duration and fixity that "closed in on itself," these placeless places intensified the disaster's disruptions. No longer "places of adventure [and] reserves of imagination," these heterotopias imploded the urban anxiety of a society that might have prepared to "go camping" but had ended up struggling for survival.[57]

In the days following Hurricane Katrina, FEMA provided an array of housing solutions, from platting large-scale trailer developments to leasing trailers for specific sites. In the first case, contracted plans provided the basis for new trailer cities, typically outside of metropolitan areas. FEMA contracted with Fluor, Bechtel, and the Shaw Group for the planning and layout of these developments. Laid out according to these plans and categorized as a "temporary group housing site," Renaissance Village, near Baker in East Baton Rouge Parish, has a population of almost two thousand hurricane evacuees.[58] Given contemporary scales of production, these "instant cities" have become possible responses to postdisaster housing problems, but their normative design disallows communal interaction and placement within an urban context. Difficulties of mobilization and political forces also slow their deployment—a reality imaged in the aerial photographs of undelivered and stockpiled FEMA trailers in Hope, Arkansas.[59] FEMA has been criticized for its security measures[60] and inflexibility at Renaissance Village—exemplified in its initially disallowing a communal tent structure at the center of the village. More broadly, the typology of "FEMA City" has emerged as the designation of the agency's postdisaster housing sites. Permutations of FEMA City include the rooted settlements that have remained for more than two years in Florida's Charlotte County following the 2004 hurricane season.[61] The "instant village" also does not include the "camping scene"[62]—meaning that these cities, when they do become "places," are either imaged as nostalgic efforts of re-creating neighborhood life or are scandalized as territories of poverty and despair, framed through the media's scenographic apparatus.

The postdisaster camp is also a dispersed network of individually leased sites. FEMA Form 90–96 administers a "Temporary Housing Pad Lease." Negotiated by the FEMA field representative, this formulary structures the agreement between the owner of a site and a disaster victim. Its agreement stipulates that the "initial term of the lease should be for the shortest possible period that can be negotiated with the Owner/Agent and shall not exceed 90 days without authorization." The pairing of trailer and house redefines the home-place as a ground, whether by ownership or by lease, on which the process of remaking both house and home is carried out.[63] Just as the fixed slabs anchor the recreational camping at Slab City, the pads offer temporary grounding in postdisaster displacements.[64] Like FEMA City, the concrete pad becomes a residual part of the postdisaster landscape. FEMA restricts the duration of the trailer cities by limiting housing assistance to eighteen months.[65] Extensions for Port Charlotte's FEMA City have allowed residents to remain since the 2004 hurricane season, and the recent construction of the community area at Renaissance Village calls into question the provisional nature of these trailer sites. The completion and dedication of the Renaissance Village Community Plaza in October 2006 underscores the educational and social needs generated by the trailer city's longevity. The plaza is designed, however, so that it can be dismantled and moved to other FEMA sites.[66]

After a disaster, the camp is typically a necessary form of temporary housing and relief operations. Its immediacy and flexibility make the camping form a useful model for postdisaster responses, but camp form does not resist permanence. A predetermined transience, ninety days with a pad lease and eighteen months with a trailer lease, is fundamental to FEMA's deployment of "temporary group housing sites." At the same time, the inherent fixity of the mobile home park is a confirmed reality of our built environment. The question that remains after the New Orleans response to Hurricane Katrina in particular and to U.S. disasters in general is how to mediate the rapidity of response, the flexibility of housing solutions, and the maintenance of citizens' connections to their home-place.

A few months after the disaster, FEMA proposed strategies of urban infill to provide housing for New Orleans residents. Sites for these housing clusters included vacant lots, playgrounds, recreational fields, parking lots, and selected areas of New Orleans City Park.[67] For the most part, these planned zones occurred within the city's limits and in many cases were close to the inner-city districts most devastated by the hurricane's impact. These locally asymptotic housing groups met with resistance from politicians and other residents of the city. In one instance, not unlike a preemptive NIMBY-ism, the juxtaposition of camp and preexisting community elicited a cease-and-desist order from New Orleans Mayor Ray Nagin.[68] Proposals for the City Park placed trailer communities on the southern edge of the green space, adjacent to an already fragmented urban edge.[69] The advantages to these urban camps were their proximity to resources, their connection to the places with which residents identified, and their acknowledgment of some degree of permanence within the slow rebuilding process.[70] Controversial political decisions generated ad hoc urban camps established by displaced groups of citizens. Residents of St. Bernard Housing Development formed Survivor's Village, a tent city in the 3800 block of St. Bernard Avenue, to protest HUD's closure of public housing. Many of FEMA's proposed interventions were disallowed by the agency's own flood zones restricting low-level housing, and other camps within New Orleans' urban context remain distinct "cities within a city." But the urban strategy of infill has revealed the possibility of a more labile housing solution within an infrastructure of functions (transportation, health care, education) and identities of place that are already well-established and in the process of being reaffirmed.

The fugitive forms of postdisaster camping urbanism require a connection to place. The occupation of the Superdome and the Astrodome for eight days underscored the problematics of large-scale responses that do not in some way address place. Such forms of response cannot invent place through the imagery of a unit's design, and the camping forms themselves cannot remain at a temporal and physical distance from the urban experience. At the same time, it is naïve to think that camping environments can be fully integrated into a preexisting urban context. Such results require an organic invention of place fostered by political fissures as in Gibsonton's midway-homes or in the small-scale, idiosyncratic camps of necessity created by the Tin Can Tourists. The municipal camps of the 1920s and 1930s have already shown the difficulties of protracted urban camping.[71] But the reverberations of furnishing postdisaster housing inevitably will extend beyond the eighteen months allowed for by FEMA. The acknowledgment of the asymptotic nature of camp's fugitive form accommodates a proximity and a necessary distance, a provisional permanence rather than a planned temporality—which is the planned obsolescence of FEMA City's temporary communities. The asymptotic touches without intersecting, and, like the platforms of Manila Village—their traces not more than thirty miles to the south—evidences the efficacy of this model of duration and freedom. Not a city in a city as enclave, but a city *on, over, before,* . . . a city, not metaphorical in its real experiences and reconnections, but humanizing in its contingencies and its inherent status of waiting. Not waiting for meaning or for funds, but for transition that is sure to come. The asymptotic postdisaster camp allows for the necessary hesitation at departure, before breaking back into the habitual and recognizable domestic spaces of home.

PART III
RETHINKING

February 10, 2006 (Fusina, Italy)

Before the Savannah camping trip, I had traveled through the Veneto's Mestre district to study the camp designed by Italian architect Carlo Scarpa. In this region, recreation camps are shadowed by heavy industry and its manufacturing presence—a consequence both of northern Italy's densely built-up environment and of the layered histories so close to the quintessentially stratified site of Venice. (I would be reminded of this juxtaposition two months later at the Savannah camp.) Scarpa's camp, positioned on the spit of land at Fusina, provides an extraordinary vista of the Venetian Island—allowing us to read its fabric at a distance of three miles. Facing this view with the Adriatic wind in your face, you might forget the looming presence of Centrale Andrea Palladio no farther than a thousand feet to the north.

In the camp's original layout, Scarpa referenced the perspectival connection to the storied monuments of Venice—one of the camp's main avenues aligns with the Cemetery of San Michele. With his design for the camp, Scarpa is thinking about, even meditating upon, his own relation to Venice. At a necessary distance, he is gathering the place of his birth. The architect had grown up near Vicenza, in a sense the Veneto's political periphery. While he had certainly worked in Venice prior to the Fusina commission and was by all accounts a Venetian architect, we might speculate that Fusina was his preparation to renew his homestead at the cusp of his career's zenith. The Olivetti showroom and Correr museum projects immediately followed, and three years later Scarpa would work on the Stampalia. Pushed further, Scarpa was rethinking how to approach Venice—he was recalling dwelling (through the temporal grounds of the camp) and reinventing his home-place, through the distance and proximity afforded by drawing and thinking. Like Dante, abiding in the linguistic home of his new vernacular and looking back to his old home from the Florentine hills, Scarpa's hesitation between arrival and departure afforded a productive camping home. Scarpa's camp at Fusina extends memory and contracts experience.

Scarpa is also thinking with his hand, in the exercise of drawing. On one sheet of paper (330 x 594 mm), a singular "ground," Scarpa thinks and rethinks, makes and remakes, the design of the camp. The drawing itself becomes a process of siting and clearing a campsite through time. He in effect thinks the multiple—delaminating the site's many grounds and proposing simultaneous experiences from many perspectives. I did not technically camp at Fusina, but I found that my fieldwork there was a kind of camping—not unlike Scarpa's own activity of obsessively redrawing the site. The graphite residue tracing previous occupations, Scarpa thinks (and builds) the site, preparing us to dwell within its poetic location. The Fusina camp is a departure point for the architect's return home.

Chapter 11

BREAKING CAMP

These rules do not trace a method, but very precisely an exodus, a capricious and seemingly irregular trek constrained only by the obligation to avoid speculative places held by force.

—MICHEL SERRES, *The Troubadour of Knowledge*

From Permanent to Temporary

Schuman says the best plan for Bridgeville is to tear it down, but he concedes it might make a good summer recreation area, perhaps an RV campsite.

—Associated Press, "Online Bidding for Declining Town Hits $357,000" (December 25, 2002)

Breaking reverses the often ineluctable movement from the temporary to the permanent. With the exodical shift, situatedness and permanence are again returned to a transience. Historically, provisional camps have yielded enduring cities—Istanbul has been described as an early encampment by Michel Butor, Johannesburg began as a mining camp, and many European cities such as York, Manchester, and Vienna have their origins in the Roman legionary camps called *castra*.[1] Generated from campsites particularly in the temperate climate of the southern United States, trailer and mobile home parks have similarly reached a degree of permanency that contradicts the transience associated with such clustering of "mobile homes." Natural disasters redouble these ironies, as the permanence of cities such as New

Orleans is undermined and as officials resist the intransience of mobilized housing. In recent years, however, hurricanes have not been the only forces to render cities and established communities temporary, and the ad hoc and paradoxical stability of "mobile homes" is being reversed as the latent ironies of portable housing are destabilized by a new set of market forces.

Beginning in November 2002, the sale of the town of Bridgeville, California, on eBay further complicates the classification of camps. The proposition that the town become a recreational vehicle (RV) campsite demonstrates not only a movement from public to private but also a departure from the expected permanence of cities and towns to the more temporary occupations of a camp. Ironies are intensified: mobile homes return to the road, cities become camps, and one moves both "from campsite to campsite" and "from camp(site) to campsite."[2]

> Although I am no archeologist, I love Florida as much for the remains of her unfinished cities as for the bright cabanas on her beaches. I love to prowl the dead sidewalks that run off into the live jungle, under the broiling sun of noon, where the cabbage palms throw their spiny shade across the stillborn streets and the creepers bind old curbstones in a fierce sensual embrace.
>
> —E. B. WHITE, "Upon a Florida Key"

With the rise of real estate prices in previously marginalized areas, mobile home parks are being sold to provide land for higher-end developments. In southern Florida, many of these parks have existed for over thirty years, and more than half of all mobile home parks in Florida were established before 1970. In southern Florida, mobile homes form a "precarious niche in the housing system" and are becoming increasingly temporary as the real estate market changes.[3] In January 2005, residents of Gulfstream Trailer Park in the Florida Keys were evicted after two years of lawsuits. In the three-year period between 2000 and 2003, three thousand mobile homes were resold and removed from their homesites. And the trend has continued with the sale of Fiesta Key's KOA campground to Cortex Resort Living in October 2006.[4]

Such a paradoxical process is not characterized by demolition or eradication but by a more subtle, incomplete transformation that blends making and unmaking. Camping procedure itself is evoked: siting, clearing, making, breaking, siting, clearing . . . Residents of the site do not typically own the land of the mobile home park, but they often do own the dwelling units, whether mobile homes or travel trailers. While the former poses greater difficulties of mobilization than the latter, both types of vehicles are salvaged and relocated to new sites outside of areas targeted by new developments. And the inherent vulnerability of building sites in the Florida Keys makes any new construction a tenuous proposition. Departure remains imminent, or at least probable—whether to escape a hurricane's path or to make room for a subsequent tenant or development. This collision of economic forces, housing necessities, and dislocated identities of home occurs within campsites of contention. These zones refocus the conversation about place to include the ground of sitedness as well as the regional domain of political dialogue and social disparity.[5] Not surprisingly at the margins of region, state, and nation, these permanent campsites of the Florida Keys, returned to a temporary status before the execution of planned developments, frame contemporary and future problematics of home—whether postdisaster, postcapitalist, or simply posturban.[6]

From camp(site) to campsite and then from campsite to campsite, we follow this constantly shifting ground that can in the end be classified as neither temporary nor permanent. Remembering that Nietzsche wondered if there could be a grounding without ground, we might ask the following: does the confluence of a contemporary itinerancy of American dwelling (from permanent to temporary)[7] and the incidence of camps that linger as dwelling sites (from temporary to permanent) suggest an alternative method for the study and construction of place—one that embraces the paradoxes inherent in architectures of mobility and time? In the Kings Road house, architect R. M. Schindler responded to this paradoxical situation of chronic itinerancy and material inveteracy, arising out of hesitation at departure.

Beware, o wanderer, the road is walking too.

RAINER MARIA RILKE, quoted in Jim Harrison, *Off to the Side*[8]

[W]e're really camping,—you'd never see how folk could live at all in such a-rough and incomplete household, it's quite wondrous . . . tho [*sic*] we've been too busy even to note the dramatic moment moving into our own ought to be.

—PAULINE GIBLING SCHINDLER to unknown (May 22, 1922)

The Kings Road house occupies the paradoxical territory of the temporary and the permanent. R. M. Schindler initiated the concept for the house during a camping trip in October 1921. On their long-awaited vacation, Schindler and his wife, Pauline, camped along Tenaya Creek in the High Sierras after completion of the plans for Frank Lloyd Wright's Barnsdall House in September. The Schindlers had outfitted their secondhand Chevrolet touring car in the same year that Elon Jessup's widely distributed *Motor Camping Book* was published.[9] In his reply to Richard Neutra's letter, Schindler wrote: "I received your letter high up in the mountains where I am having a vacation for which I have waited a long time. It is one of the most marvelous places in America. I camp at the shore of the Tenaya, sleep on a bed of spruce needles under a free sky and bathe in the ice-cold waterfall." He also described his possible departure for Japan to supervise Wright's work following the completion of the Imperial Hotel, noting dramatically that "everything will be decided in the next two weeks."[10] Born in Prague, Rudolph M. Schindler had immigrated to the United States from Vienna in March 1914 aboard the *Kaiserin Auguste Viktoria*. Before the camping trip in June 1921, Schindler had written Neutra from Los Angeles that he was "still not at home here" and that he was considering a return to Vienna.[11] Having subsequently ruled out this return to Vienna, Schindler decided that, if work in Japan were to fail, he would remain in Los Angeles and "build a small studio" from which to establish his architectural practice.

A house roams at night when its occupants sleep.

JOHN HEJDUK, *Such Places as Memory*

While on the camping trip, Schindler began the design for the Kings Road house and immediately after his return completed working drawings for the project. On November 1921, Schindler finalized a preliminary plan for the house, with red poché walls forming U-shaped spaces within the three-part pinwheel plan. Property in West Hollywood at 835 North Kings Road was purchased later that year; and after only a few months of building activity in May 1922, the Schindlers along with Clyde and Marian Chace moved into the unfinished house that had been granted a temporary building permit because of its untested techniques of construction.[12] Having recently completed a treatise on the pleasures of camping, Pauline Schindler described living at the Kings Road house as an extended camping excursion within a chronically unfinished household. Cast as an experiment in cooperative dwelling, Kings Road summarizes the architect's concept of house as a permanent camp.

> The house likes the weaver; it remembers its early construction.
> JOHN HEJDUK, *Such Places as Memory*

Schindler describes the architectural scheme of the Kings Road house: "Each room in the house represents a variation on one structural and architectural theme. This theme fulfills the basic requirements for a camper's shelter: a protected back, an open front, a fireplace and a roof."[13] Schindler calls the house "a simple weave of a few structural materials" and notes the "organic fabric of the building." Although referring to the house as a primitive cave in other writings,[14] Schindler here makes explicit reference to the fabric walls of tents and draws a connection to Gottfried Semper's thesis on the "dressing" of architecture.[15] The panels of the tilt-wall construction can be read as pieces of fabric in the immediacy of their creation and the "lightness" and facility of their assembly. The formwork for each slab requires that a three-inch space be left between the wall units. In some cases, Schindler has glazed this space, which then becomes an open joint between the walls "to filter air and light"—effectively dematerializing the heaviness and confinement of the concrete walls. These partings of the wall correspond to the parted flap of the opening in Schindler's tent. Canvas walls and sliding panels and doors provide tentlike enclosure along the open ends of the U-shaped spaces. If the sleeping baskets are a smaller version of the spaces of the house, then their canvas walls and roofs also point toward the tent as shelter. Although ascribed to the economy of material, the camber of the slab walls toward the top also alludes to the taper of a tent awning. Echoing this interest in the tent as shelter, Pauline Schindler will write the unpublished manuscript, "Joys of Tent Life in California" upon their return from the camping trip in the High Sierras and during Schindler's work on the design and construction of the Kings Road house.

Other aspects of the Kings Road house that relate to camping and the

R. M. Schindler, "Our Tent," Yosemite National Park, Camp 9, October–November 1921. R. M. Schindler Collection, Architecture and Design Collection, University Art Museum, University of California, Santa Barbara.

R. M. Schindler, Kings Road House, Perspective exterior elevation, 1921. Graphite on tracing paper. R. M. Schindler Collection, Architecture and Design Collection, University Art Museum, University of California, Santa Barbara.

campground are the fireplaces, the treatment of figure-ground, and the idea of collective living. Each of the six living areas (including the "outdoor rooms" of the terrace and court) has an open fireplace. With the requisite "open front" and "protected back," the openness of each of these spaces transforms the domestic fire into campfire. As a figure-ground composition, the Kings Road building site, like the campsite, integrates and mixes interior and exterior. The house shapes the site and the site shapes the house. As with the dominant ground in camp's figure-ground, the house's ground of the courtyard spaces molds the flexible enclosure of the partitions and edges. The figure of the walls remains the porous membrane defined by the reflected light of the "campfire" and fixed only by the fireplaces and the position within the overall pinwheel organization of the plan.

Prefiguring his experiment with communal and multiple housing in the 1930s, Schindler's design of the Kings Road house framed an attempt at collective living. The Chaces joined the Schindlers for the first two years, followed by Richard and Dione Neutra and a succession of other couples and friends. The shared kitchen, bath, and dining room form the public core of the house, while the peripheral spaces allow for each couple's semiprivacy. Like the intimacy and relatively open living found in many camps, the Kings Road house was an experiment in communal living, made possible through its design. Schindler would revisit this idea in later projects, particularly the Beach Colony for A. E. Rose in Santa Monica, California.[16]

Hesitating

> I came to live and work in California. I camped under the open sky, in the redwoods, on the beach, the foothills and the desert. I tested its adobe, its granite and its sky. And out of a carefully built up conception of how the human being could grow roots in this soil—unique and delightful—I built my house.
>
> —R. M. SCHINDLER to Esther McCoy (February 1952)

Hesitating prior to his departures for Japan and Vienna, Schindler maintains camp in Yosemite, at Kings Road, and throughout the California region. At home in the bricoleur's laboratory, the architect does not hesitate out of indecision but lingers, waits, and ultimately submits to the forces of matter

Construction view of Kings Road House—“making camp” and raising the “slab-tilt” walls, 1921. R. M. Schindler Collection, Architecture and Design Collection, University Art Museum, University of California, Santa Barbara.

Running-board combination bed and tent. Elon Jessup, *The Motor Camping Book*, G. P. Putnam’s Sons, 1921.

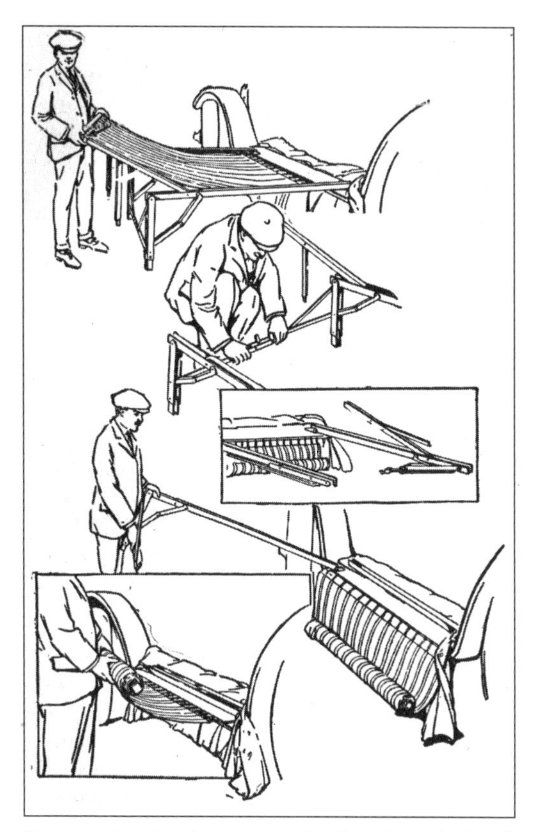

Diagram showing the process of taking down the Schilling steel bed and tent. Elon Jessup, *The Motor Camping Book,* G. P. Putnam's Sons, 1921.

and place. The siting of this paradoxical combination of Virilio's speed and Kundera's slowness occurs within the camp, within such constructions as Kings Road. Deleuze's rhizome at first appears to oppose Schindler's "growing roots," but the idea that place acts on the nomad-architect remains.[17] Schindler's immigration-homelessness-nomadism does not end with his decision not to break camp in California. His nomadic spirit is temporarily localized. And in Schindler's case, the camp is both vehicle and process, a "grounding without ground." The Kings Road house serves as a prototype for a flexible "traveling-in-place" that corresponds to his understanding of

Wedge-style tent attached to automobile: "With a shelter of this type you can build the camp fire out front and sit in the tent facing it." Elon Jessup, *The Motor Camping Book,* G. P. Putnam's Sons, 1921.

R. M. Schindler, A. Gisela Bennati Cabin, exterior view, Lake Arrowhead, California, 1934–37. R. M. Schindler Collection, Architecture and Design Collection, University Art Museum, University of California, Santa Barbara.

contemporary notions of home and dwelling: "the modern dwelling will not freeze temporary whims of [the] owner."[18] Also, the house attracts other travelers—artists and luminaries visit the Schindlers as the house becomes a fixed zone of contact among fellow "campers." Campers who occupied the Chace studio at Kings Road included John Cage and his friend Don Sample, who lived there for nine days between December 31, 1933, and January 9, 1934.[19] Cage writes of the Kings Road house, "The entire effect was one of horizontality and a sort of organic calm which was not exciting but rather full and complete."[20]

Casting Off: An Interlude between μετα + ο'δο' and ε'χ + ο'δο'ς

Mixing episodic experiences of place with Bergsonian duration, the camp(site) allows for a coexistence (however uneasy) of the temporary and the permanent. Camp as method (μετα + ο'δο'ς), reworked as an exodical (ε'χ + ο'δο'ς) movement, yields the possible simultaneity of untying and retying. Casting off, after breaking camp, returns the camper to movement: "When seafaring craft cast off, they turn their sailyards toward a world that is strange beside the landlocked daily routine: on the plain of the high seas, nothing ever resembles what's been left behind. What's square becomes round, what's stable moves; you'll never make the same movements, you'll speak a singular language, which no one who hasn't been there will understand. To leave is to sever all bonds. To go out from this world and enter another, where nothing will be the same: that's called casting off. Equipped with their gear, foreign to land and adapted to the sea, loosing their hawsers and cutting fabric of former connections, vessels are capable of providing this shattering transition. We're going to live differently, perhaps for a long while, elsewhere, where the watchman will have only the wind and sky for companions; that's why sailors always have about them, when they return, that odd little air."[21] Tied up in this renewed itinerancy of departure is an assumed arrival. In its "unsiting," breaking retains elements of resiting. *Einbruch*, a breaking into space, permits the opening, the Heideggerian lightening of specific places. And space is cleared for "diverse engagements."[22] Breaking allows for the possibility of invention and is always acknowledged in the creation of "camp space." Through the process of casting off, breaking in effect relinks and refastens camp to place. Camping cannot be romanticized as method or place. The recreational campsite maintains that "odd little air" of new and possible homes that travels with the dweller. On the frontier of different places and a time that draws the lessons of the past and the vitality of the future to a present home.[23] Hesitating at departure.

Chapter 12

DEPARTING CAMP

Meditations before Returning Home

When my floor was dirty, I rose early, and, setting all my furniture out of doors on the grass, bed and bedstead making but one budget, dashed water on the floor, and sprinkled white sand from the pond on it. . . . It was pleasant to see my whole household effects out on the grass, making a little pile like a gypsy's pack, and my three-legged table, from which I did not remove the books and pen and ink, standing amid the pines and hickories. They seemed glad to get out themselves, and as if unwilling to be brought in. I was sometimes tempted to stretch an awning over them and take my seat there. It was worth the while to see the sun shine on these things, and hear the free wind blow on them; so much more interesting most familiar objects look out of doors than in the house.

—HENRY DAVID THOREAU, *Walden*

Walden as Camp

Thoreau went to the woods, and we go camping. His site at Walden tied together place and time with an experiment in deliberate living. When he must clean his cabin, he relocates his interior furnishings outdoors. Thoreau repeats this activity and occasionally considers leaving the externalized domestic assemblage intact with an improvised awning. This hesitation reflects our own contemporary pause at confluences of interior and exterior and of temporary and permanent. We often find home while we are away from home. Sometimes upon returning home, and at other times by dislocating our own conventions of domestic life. Thoreau is camping at home.

If breaking camp makes room for leeway in the production of space, then departure returns the inhabitant to movement and at the same time refastens camp to place. Each campsite is a network of related sites, experiences, and ideas about camping. After breaking camp at Yosemite, the architect Schindler translates tent flaps into the tilt walls of his own house and, neither relocating to Japan nor returning to Austria, ties his new California home to a professional break with mentor Wright. Thoreau also considers a multiplicity of sites through his own experiment in living. Finding the site for his house requires an attitude between anticipation and transition: "At a certain season of our life we are accustomed to consider every spot as the possible site for a house."[1] This openness to all sites is not unlike the Janus-like moment of breaking and departing camp, in which we project ourselves to the occupation of future sites,[2] while recalling (sometimes with an unexpected nostalgia) the recent occupation of our current site. Place, then, is the region that holds these campsites and is also the pragmatic conditions of the individual campsite. Place is the extended zone of itinerancy for the Tin Can Tourists and Gibsonton's carnival performers. It is also Thoreau's many possible crystallizations of the tent as house during his wide-ranging hiking excursions in the Northeast. At the same time, place is the deliberate life of negotiating the site itself. Manila Village's place is a pragmatic outcome of the tide's fluctuations in the Delta region. Slab City appropriates the foundational concrete relics from a previous encampment, and Gibsonton emerges from the political fissures of zoning regulations.

Campsite

> . . . every spot as the possible site for a house . . . wherever I sat there I might live, and the landscape radiated from me accordingly . . .
>
> —HENRY DAVID THOREAU, *Walden*

Breaking ties camp to its immediate place, and departure engages place as region. In the process of camping, the region itself becomes a place. A multiplicity of campsites makes up the resulting place-as-region, which accommodates simultaneities of here and there, near and far, large and small, global and local. The territory can be found in the detail, and the detail in the territory. Such places are thus not delimited either by locating or partitioning.[3] Positioning of place might occur at a distance, and region can be understood as place if the fluidity of time and movement is considered. For the performers living in Gibsonton, the seasonal network of carnivals and fairs serves as the region in which these particular reconstructions of the midway occur. The construction of the midway formulates the local operations that characterize the relation to place. Though arising from the "model of the midway," the flexibility and fluidity of local operations allow for each place to be understood in its specificity and in its connotation as region. *Any* place in the region becomes *some* place through the construction and performance of the midway in Gibsonton. The ritual of dancing on the platforms of Manila Village locates its diverse community of shrimpers, and the annually generated campsite of Burning Man draws together the society of "burners," joined for the rest of the year through the electronic regions of the virtual camp.[4]

The campsite is also a pragmatic place. Place-as-pragmatic is a milieu, or middle ground, in which something is made. In terms of dwelling, this condition is not a simple positioning along a surface nor is it a dwelling through complete rootedness, or depth. Instead, place-as-pragmatic is place as process in which operations, as "local operations," define and are defined by attributes of place. To dwell lightly, as in the implementation of the midway and the development of the town of Gibsonton, is to press the ground but to do so with respect to configurations of local details and to the larger territorial implications. Taking into account William James's call to ambulation, place-as-pragmatic occurs between location and fixity in a less structured, though equally rigorous, movement "hither and thither."[5] It refers to the place where things are worked on,[6] where the ground is molded, scraped, and pressed, and where formal ordering is derived from imperfections in the site—requiring a series of operations, siting, leveling, stacking, laying, and pressing. Place-as-pragmatic meets place-as-region at the confluence of these local operations of camping.

Place can also be understood as a negotiation between the local absolute and the absolute local. And camps are paradoxical places that lie at the *chi*, crossing, and mixing of these two conditions. Analogs for these poles might be the ship and the colony. Similar to the detail described above, the Greek ship, as the mobile unit at sea and as an immobilized, turned-over vessel at seaside camps, is the local absolute.[7] This form has been concretized in

Italian, Irish, and Norwegian keel churches. On the other hand, the Roman colony-*castrum* can be understood as the absolute local, its gridded matrix having been imposed on and in the ground and thus defining a colonial territory that lies at the regional extreme.[8] The local absolute is not the opposite of the absolute local.[9] As places of mediation and paradox, camps often include attributes of both the local absolute and the absolute local, and tension between these factors characterizes the process of camping.

In Slab City, the motor homes offer a potentially infinite "succession of local operations" that can be considered as the initial stages of the local absolute. The slabs themselves contribute the condition of the absolute local—not because of their fixity alone, but primarily in their having been overlaid and pressed into the ground as a regulating apparatus. These open platforms are limiting but they no longer completely delimit. The slabs now float on the sea of porous, "superstitious" clay and depend on external forces and extra-ordinary occupations for their definition. This indefinite quality allows their incorporation *into*, as opposed to their previous definition *of*, a particular place. Through this combination of localization and indefinitely limiting forces at Slab City, local becomes absolute for a particular duration. The internally generated event of Slab City is "grounded" through the slabs. And place can then be constructed through this event. From the understanding of place-as-region, we can derive the idea that place is everywhere. This wide-ranging potential for the construction of place does not point toward a lack of specificity of place. Instead, place is made manifest by way of particular actions—the local operations. Such places as Slab City and Gibsonton resist classification solely in terms of space and must be understood in terms of movement and time—actual mobility and historical flux.

The campsite becomes an event-place. Each campsite is inherently a set of related camps and ideas about camp. Each set is constructed from previous sitings on or near the actual campsite, from historically related camps or phenomena that resonate with situations of camp, and from the notion of camp as an idea or method closely related to each case study's particular place-situation. The historical layers of campsite as place provide a simultaneity in their residual siting. Ruskin is sited over and within a preexisting turpentine camp. Gibsonton is sited adjacent to a fish camp. The platform camps of Manila Village are sited near the fish camps and houseboat communities of Barataria Bay and the Atchafalaya Swamp.[10] Slab City is sited over a series of Indian camps and within a former military training camp. And, according to legend, Braden Castle Park is sited over an Indian camp, which afforded access to the medicinal qualities of a natural spring and of the confluent waters of the two rivers.[11]

Florida's thinly imbricated histories of place and space are narrated through its campsites.[12] Documentary photographers found these stories in the municipal camps of the Tin Can Tourists and in such camping artifacts as the shell gardens that became ad hoc "welcome mats" and defined each campsite's identity within a community of tourists. In spite of these connections with the past, the campsite remains incomplete and relies on subsequent occupations and transformations to continue its development as a place. Maintaining its partial completion, the campsite witnesses events. The camp at Braden Park preserves the memories of the castle and its landscape of events. The camps of Gibsonton register the changing and dynamic culture of the carnival performer, and Slab City is a site for understanding the practices of camping cultures ranging from established Canadian snowbirds to a shifting demographic of exurban squatters. With the event, "campsite" also becomes "camp(site)," its sitedness partially suppressed, and in some ways intensified, by temporal exigencies of itinerancy.

Camp(site)

> As for a habitat, if I were not permitted still to squat, I might purchase one acre at the same price for which the land I cultivated was sold. . . . But as it was, I considered that I enhanced the value of the land by squatting on it.
>
> —HENRY DAVID THOREAU, *Walden*

Camps and campsites are understood through time, specifically a time of duration. The second theme introduced in the opening of this work noted the idea of a temporary presence in the process of becoming permanent, or reaching a degree of permanency. This degree of permanency depends on time rather than space. Action, *in* and *through* time, rather than perceptions *of* space defines the transience of the camp. Camps are then Bergson's "zones of indetermination." Duration occurs in the middle temporal ground of the camp(site) between camp and campsite. For Bergson, duration is a "way of being" that is partially revealed in its process. In a statement that resonates with the discussion of the local absolute, Deleuze notes that for Bergson there are two sides to any location or absolute position: "Duration is always the location and the environment of differences in kind; it is even their totality and multiplicity. There are no differences in kind except in duration—while space is nothing other than the location, the environment, the totality of differences in degree."[13] The "way of being" is thus a combination of location and environment, the *lieu* and the *milieu*. In Bergson's model, "differences in kind" allow for a true multiplicity of experience.[14] Duration is the composite, aggregate, and environment for this type of difference. Bergson thus helps to reframe the problem of uncritically equating space and place. While his focus on time as opposed to space and his equivalence of space and location obscure the role of place in our experience of the world, our "way of being," Bergson's redefinition of space through time does suggest a type of place related to the subject of this study. Deleuze's explication of this Bergsonian "new space" resonates deeply with this discussion of place: "If things endure, or if there is duration in things, the question of space will need to be reassessed on new foundations. For space will no longer simply be a form of exteriority, a sort of screen that denatures duration, and impurity that comes to disturb the pure, a relative that is opposed to the absolute: Space itself will need to be based in things, in relations between things and between durations, to belong itself to the absolute, to have its own 'purity.' This was to be the double progression of Bergsonian philosophy."[15]

This new space will be found within, and come from, things-in-the-making—a procedure related to the idea of place-as-pragmatic. Place is not simply remolded or remodeled "matter" but a composition of differences of kind and memories whose origins differ absolutely in kind. Bergson helps clarify the distinction of place and space in which camping occurs. Camps must differ in kind because the situation of camping changes with each iteration of place. Camps and their methodology remain "in the making"—momentarily lodged between the "thingness" of places and the times of camping. Ultimately, camping becomes a mediation of "between things and between durations" and in doing so becomes a meditation on place.

To comprehend the role of place in cultures and conditions of itinerancy, place can be conceived as a negotiation between detail and territory through time. In this model, the detail serves as a locus for the incorporation of external attributes. Such loci include the mobile detail of the camping vehicle and the temporary structure of the camping shelter. This exchange returns to a theme introduced at the outset of this work, that of the exteriority incorporated from within, and thus occurs in movement and in partial fixity, in the making and in the made, and in the series of local operations and in their previous permutations elsewhere. The reality of these situations is their transience such that we must resort to Bergson's "things in the making" as opposed to "things made" to understand how such places are constructed. Manila Village, Gibsonton, the municipal camps of the Tin Can Tourists, Slab City, and the festival of Burning Man are all unfinished constructions. Their completion always awaits the next season, the subsequent camping tour, or the tidal ebb and flow.

We should not forget that Thoreau was squatting at Walden. He moves his belongings outside from within a house that he does not own. This double dislocation is the camping life. Place is not always emplacement, and time is neither fleeting nor unchanged. Squatting has an unknown duration, and these paradoxes of place and time define the camp as the camping home.

Camp

> Both place and time were changed, and I dwelt nearer to those parts of the universe and to those eras in history which had most attracted me. . . . I discovered that my house actually had its site in such a withdrawn, but forever new and unprofaned, part of the universe.
>
> —HENRY DAVID THOREAU, *Walden*

Camp as place. Camp as duration. Camp as method. Camp as idea of home. Within the places of campsite and the fragile duration of camp(site), we also find camp as home. The confluence of place and time in camps yields a productive destabilization of home that suggests new methods and ideas about working within places of itinerancy.[16]

The camp, particularly the camp as carnival midway, becomes a broken threshold.[17] Just as camps are mobile sites for dialogues with place, the broken threshold is an "artful connection" that reties detached elements or disconnected itinerants. As an example of place-making at home, Penelope works within the broken threshold as she at one and the same time constructs the route and traverses the itinerary of Ulysses.[18] In making and traversing, Penelope weaves the places of travel from a given location. Her woven cloth narrates Ulysses' movement from place to place, from campsite to campsite, only to be repeated with the breaking of each camp.[19] Place in this respect is a middle ground between a totalizing local and an encompassing global.[20]

And home is the site for gathering, and sometimes weaving, these disparate places together. As a vernacular construction, the camp incorporates the external *from within*. The broken threshold resists departure and asks that the dweller hesitate. This threshold, like the method of thresholding described at the outset, occupies a middle ground between detail and territory—the indeterminate zone from which static buildings emerge but in which such tendencies are temporarily suppressed by the objectives of a camping life. The appended "from within" characterizes the quality of "indwelling" or inhabiting a place deeply that has always been a part of the vernacular's meaning and that has resulted in the singularity of its usage. Yet this in-habitation is very important to the overall idea of the internal manifestation of outside forces. The site for this expression of the external is home.[21] Camps are also homes that reconstruct place through the realities of the situation and memories of the home that has been left behind.[22] Camp as home internalizes the positioning of site and the passage of time. And as a home-away-from-home, camp is closely related to the idea of home. Thoreau's home and his idea of home are firmly intertwined so that he can write of his revelation that his "house actually had its site in such a withdrawn, but forever new . . . part of the universe."

As an act of distancing and maintaining propinquity, the camp becomes a unique coincidence of two homes. One home is a fixed point of reference, whether childhood home, family homestead, or simply a "permanent address," and the other "home away from home," which can be a vacation destination, seasonal tour, or weekend retreat. One way I have tried to understand this possible simultaneity of homes is through the idea of the "nomad at home." In this concept, home is not a singular place or idea but instead must embrace the multiplicity of recollections and resonances *within* and *around* home. Through travel and inhabitation, the nomad circumscribes and inscribes the construction of home, which then becomes bell hooks's place of discovery—a ground made productively unstable by traveling *in* place. Home thus combines the rhetorical territory and the Bachelardian garret—or to return to the introduction, the memory and mental image of the *mattang* and the Airstream trailer.[23] Camps, as reconstructions of home, also negotiate this zone between distant homes, as both the physical construct of the house and the mental construct of home, and the specificity of the "new place" of the campground, which is hopefully, like eutopia, a good place. Camps become places where the idea of home is recast both in actual construction and in the reconstruction of the home-place through stories.[24]

Camp as Method: Reading Virgil in Tents

> We read Virgil and Wordsworth in our tent, with new pleasure there, while waiting for a clearer atmosphere.
>
> —HENRY DAVID THOREAU, "A Walk to Wachusett"[25]

In August 1858, William James Stillman guided a group of philosophers through the mountainous terrain of the Adirondacks to Camp Maple. Having recently returned from England and a painting mentorship with John Ruskin, Stillman planned to document life at the camp. During this time, an emergent American tradition of camping was attracting writers, philosophers, and scientists to remote sites, away from their obligations in Providence, New Haven, and Boston.[26] Prepared as a study for a larger painting that was never completed, *The Philosophers' Camp* includes the figures of James Russell Lowell, Ralph Waldo Emerson, Louis Agassiz, and John Holmes.[27]

A significant documentation of scholarly camping and the Transcendentalist tradition at Camp Maple, the painting also illustrates camp as method for negotiating place, for thinking about and making new places. At the left, the scientists gather around Agassiz, who dissects a fish; and to the right, Lowell anchors the group of writers whose attention has been directed toward a distant valley. The darker figure of Emerson follows the writers' gaze but remains as a solitary outline between the groups. Within the grounds of the camp, Emerson is the compositional and philosophical joint. Horizontally, Emerson links science to the arts—Agassiz gathered many of his

William James Stillman, *The Philosopher's Camp in the Adirondacks,* 1858. Courtesy of Concord Free Public Library.

specimens for analysis from the northeastern wilderness with the help of Thoreau and others, and Lowell and his fellow literary travelers were poetically inspired by the rugged terrain and its phenomena. As *imago mundi,* Emerson expands the camping territory to make broader connections between nature and meditations, analysis and poetics, and experimenting and writing. Emerson's figure also works vertically, defining the center of the camp along with the dramatic angle of the camp's eponymous maple tree. As a mobile *axis mundi,* the poet links earth and sky. We can imagine the Emersonian vehicle, traversing the grounds of the camp, tending the campfire, and borrowing from each conversation and observation. The metronomic and peripatetic Emerson times the camp, his meditations pulling the future to the present while acknowledging the past: "Time makes a fresh start again, on for a thousand years of genius more."[28]

For the philosophers, the camp was not only a site for reconnecting with nature. Camping itself was a practice to generate new ideas and to make new places. In his own camping excursions, Thoreau set out with the objective to live deliberately.[29] Along with Walden as camp, the philosophers' camp was a place that permitted confluences of ground and sky, movement and fixity, and thinking and making. It was a laboratory for experimentation and speculation; the camp was something where and on which one could meditate. Later in the nineteenth century, another Adirondack site would host early experiments in the pragmatic method. In the mid-1870s, William James along with a group of Boston doctors purchased land in Keene Valley, New York, for what became known as Putnam Camp. James's essay "What Pragmatism Means" is sited in a camp that is probably Putnam Camp: "Some years ago, being with a camping party in the mountains, I returned from a solitary ramble to find every one engaged in a ferocious metaphysical dispute. The corpus of the dispute was a squirrel."[30] The debate centered around whether a viewer who tries to catch sight of a squirrel, as it moves around the tree matching the human witness's radial movement and thus always staying out of sight, moves around the squirrel or not. From this anecdote and question, James moved quickly into an explanation of the pragmatic method, "to try to interpret each notion by tracing its respective practical consequences."[31] In James's method, meaning is determined by the "practical difference" assigned to the possible resolutions. And a camping pragmatism works between the necessities of place and the poetics of experience.[32] On September 16, 1909, Putnam Camp hosted Carl Jung, Sándor Ferenczi, and Sigmund Freud, who reputedly chopped wood for the campfire.[33]

For these groups, camping practice was also a return home. While the Stillman party was camping in the Adirondacks, Thoreau was continuing to document his Walden environment, measuring its topography, walking its region, and camping along its periphery.[34] Emerson visited Thoreau on August 23 to tell him stories about the camping trip and to boast of his prowess with the double-barreled gun, bought specially for the trip.[35] Emerson would then write in his poem "Adirondacs" of the camping home[36] and its connections to ways of living: "Suns haste to set, that so remoter lights / Beckon the wanderer to his vaster home." Emerson and his "fellow travelers" find in the camp a transformation of place and time that dislocates home just as it links domesticated arrivals and departures. As they prepare to "strike" camp, Emerson notes how "Under the cinders burned the fires of home." The campfire is the new hearth, one of reverie and of emplacement. And home can then be held within both actual and imagined places:

"to spiritual lessons pointed home, / And as through dreams in watches of the night, . . . Not clearly voiced, but waking a new sense / Inviting to new knowledge, one with old." The productive destabilization of home invents a new home, one closely linked to history, experience, and projections of possible future homes.[37]

Reading Virgil in tents places poet and dweller within a changing place and time. Camp is then ultimately a method for exploring the possibilities of home—a kind of field research that is not an unrelated activity or an isolated intellectual pursuit but is instead a fundamental way to experience places. And camping becomes a reflective practice of place-making.

NOTES

INTRODUCTION

1. Henri Lefebvre has studied this mix of cyclic and linear times in everyday life in his method of rhythmanalysis, detailed in "Elements of Rhythmanalysis." See particularly *Writings on Cities*, trans. Eleonore Kofman and Elizabeth Lebas (Cambridge, Mass.: Blackwell, 1996), 228–33.

2. See the entry for "gerund" in the *Oxford English Dictionary*, 2nd ed., 20 vols., prepared by J. A. Simpson and E.S.C. Weiner (New York: Oxford University Press, 1989).

3. A parallel discussion, particularly in the context of the formatting of an academic dissertation and its "coding" and formalization, could be carried out regarding the homophonic connection between "siting" and "citing." Denotatively, citing is a summoning or quoting. Connotatively, a clearer connection exists with siting. Citing is a setting in motion, a beginning that occurs in movement (from the Latin *citâre*). Siting as an event can also be traced in the *Oxford English Dictionary*'s fourth meaning of the verb "to cite" as "to bring forward an instance." Moreover, citing like the siting of camps often occurs as an annotative process on the margins or edge of the main work, or "ground."

4. Gilles Deleuze, *Negotiations*, trans. Martin Joughin (New York: Columbia University Press, 1995), 127.

5. Guy Debord, "Report on the Construction of Situations and on the International Situationist Tendency's Conditions of Organization and Action," June 1957, trans. Ken Knabb (Berkeley, Calif., Bureau of Public Secrets), http://www.bopsecrets.org/SI/report.htm. The problem of defining suitable forms for building situations remained for Debord and the rest of the Situationists throughout their investigations. The work of Constant Nieuwenhuys remains the most "architectural" of the Situationist experiments with form. See *The Activist Drawing: Retracing Situationist Architectures from Constant's New Babylon to Beyond* (2001), ed. Catherine de Zegher and Mark Wigley. It is also important to note that the phenomenon of the *situation construite* begins with the environment of household parties (attended by

Debord and the Situationists) but is soon ideologically and politically transformed to form "a part of a cumulative revolutionary chain." Simon Sadler, *The Situationist City* (Cambridge: MIT Press, 1999), 106–7.

6. The "problem" that the Situationists encountered became the contradictory nature of their theoretical (in many ways, political) and practical pursuits. Debord's polemic of a "free architecture" and his proposal of psychogeographic research as both observation and intervention results in the duality of the Situationist project. For Debord, this form of research has a "double meaning: active observation of present-day urban agglomerations and development of hypotheses on the structure of a situationist city." "Report on the Construction of Situations."

7. Anonymous, "Preliminary Problems in Constructing a Situation," 1958, trans. Ken Knabb (Berkeley, Calif., Bureau of Public Secrets), http://www.bopsecrets.org/SI/1.situations.htm. With this document, the Situationists, in addition to outlining the attributes of the constructed situations, were attempting to resolve the problem of meshing an artistic endeavor (and its "mechanical" production of ambiance) with the more communal, accessible, and experimental zones of activity. The essay also included the following problem statement: "A constructed situation must be collectively prepared and developed."

8. Historically in a philosophical context, Edward Casey speculates that the collapse of notions of space and place into position or "site" was completed in the eighteenth century following Descartes, reaching full development in Leibniz's *analysis situs.* See Edward S. Casey, *The Fate of Place* (Berkeley and Los Angeles: University of California Press, 1997), 182–83. This model of place as site denies "inherent properties ascribed to [space and place] by ancient and early modern philosophies: properties of encompassing, holding, sustaining, gathering, situating" (183). Here, Casey is talking specifically of the *analysis situs* model of Leibniz.

9. Gilles Deleuze and Felix Guattari, *Nomadology: The War Machine,* trans. Brian Massumi (New York: Semiotext[e], 1986), 54. In contrast to striated space, nomad space is "localized and [yet] not delimited." This paradoxical condition is reflected in the discussion of siting in this section where there is the possibility of a placing without emplacement—that is, a geometric freezing of siting into site.

10. Foucault points out the connection between the architectural requirements of institutions and the leveling of sites for prisons. See Michel Foucault, *Discipline and Punish,* trans. Alan Sheridan (New York: Vintage, 1979), 231–56.

11. Michel Foucault, "Other Spaces: The Principles of Heterotopia," *Lotus International* nos. 48–49 (1986): 16.

12. Foucault's statement in "Other Spaces" reflects the conflation and confusion of space and place that Casey critiques: "heterotopia is capable of juxtaposing in a single real place several spaces, several sites that are in themselves incompatible" (15).

13. Real places are understood in Foucault's model as a proliferation of sites and possible "other sites."

14. Casey, *Fate of Place,* 300–301.

15. In this case, Casey follows closely the work of Peter Eisenman and his idea of "spacing." Ibid., 318, 335.

16. "Lightening" and "event of Appropriation" are translated from Heidegger's "*ereignen* and *das Ereignis*."

17. Lightness is an ordering principle in the Open City's design: "Lightness because the way in which the constructions touch the ground . . . lets the land initiate the configuration of territory and space in both plan and section. . . . And status of lightness because there are no apparent imposed formal ordering devices that regulate the development of the constructions." Ann M. Pendleton-Jullian, *The Road Is Not a Road and the Open City, Ritoque, Chile* (Cambridge: MIT Press, 1996), 3.

18. Albert Hofstadter, introduction to *Poetry, Language, Thought*, by Martin Heidegger (New York: Harper and Row, 1975), xxi.

19. Martin Heidegger, "Addendum to the 'Origin of the Work of Art,'" in *Poetry, Language, Thought*, trans. Albert Hofstadter (New York: Harper and Row, 1975), 84; Heidegger, *Poetry, Language, Thought*, 62.

20. Martin Heidegger, "Building, Dwelling, Thinking," trans. Albert Hofstadter, in *Basic Writings* (New York: Harper and Row, 1977), 332.

21. The case of Manila Village reveals the loose structuring of a creolized community of exiles who set up platforms in a watery fissure between legal boundaries, made indefinite by river sedimentation and tidal fluctuations. In Gibsonton, the community of carnival performers enlists the county zoning overlay known as "Residential Show Business" to allow for an inversion of typical relations between the public and the private and between the performative and the residential. Throughout Florida, the Tin Can Tourists formulate a code of the camping tourist, aspiring to make connections with the municipal organization and the zones of public access in the early semiurban landscape of Florida. Braden Castle Park clears a city within a city with its own regulations that foreshadow modern neighborhood associations. Slab City cultivates a site that steadfastly remains in the public domain as a result of its original section 36 classification designating it for public educational use and as a result of its subsequent, short-lived use as a military training facility. Although they do encompass migrancy (in Manila Village) and homelessness (in Slab City), these campsites occur within a Western framework of camping. Western philosophy thus provides the main theoretical basis for understanding these camps, but connections with Eastern traditions will also be made.

22. This correlation between camping and Daoism's way comes from a discussion with William Tilson.

23. Deleuze and Guattari, *Nomadology*, 84. The authors continue their outline of attributes for the war machine of nomadology: ". . . the weapon being only a provisory means. Learning to undo things, and to undo oneself, is proper to the war machine: the 'not-doing' of the warrior, the undoing of the subject."

24. Clearing also provides a setting to review ideas associated with the particular iteration of camping in each case study.

25. The interrelationships of the case studies characterize the broader environment of camping (geographical and methodological in terms of this work's overall objective to outline a camping practice), and the specific procedures of constructing camps are described within the "making" section of each case study.

26. Henri Bergson, "Introduction to Metaphysics," in *The Creative Mind: An Introduction*

to Metaphysics, by Bergson (New York: Citadel, 1992), 188. Pertinent here is Quatremère de Quincy's idea that architecture does not make *what* it sees but looks at *how* constructions are made. Architecture thus arises out of making rather than the made.

27. Deleuze and Guattari, *Nomadology,* 38, 98.

28. Questions that the sections on making seek to answer are ones that architect Lars Lerup has also asked in his early project *Building the Unfinished: Architecture and Human Action* (Beverly Hills: Sage, 1977).

29. In a collection fittingly titled *Poetic Localities,* William J. Stillman's photographs of Adirondack camps and his painting *The Philosophers' Camp* convey this idea of camp as poem. See illustration on p. 242.

30. Greek *poiein* and *poiesis.*

31. *Travesía* is also a nautical term to describe a crosswind.

32. Pendleton-Jullian, *The Road Is Not a Road,* 46.

33. Ibid., 85–87.

34. Ibid., 143. *Poiesis* is the action or faculty of producing or doing something especially creatively. Pendleton-Jullian notes the implied emphasis on process or the "act of creativity" as opposed to the result. In her discussion of *polis,* Indra Kagis McEwen notes that the construction (as opposed to *chora*) was "allowed to appear as a surface woven by the activity of its inhabitants; with processions to sanctuaries providing linkages to the territorial edges." McEwen argues that the *polis* was "emergent" and "made" in addition to being influenced by colonization. *Socrates' Ancestor: An Essay in Architectural Beginning* (Cambridge: MIT Press, 1993), 80–81.

35. The idea of the "open city" can also work at the scale of the house or individual residence. The domestic construction as poetic act is reflected in R. M. Schindler's understanding of the Kings Road house, which combined a campsite with poetics of construction with the regional attributes of Southern California. Chapter 11 focuses on the Kings Road house.

36. In his own thinking, Michel Serres has circumscribed a space of simultaneous arrival and departure through a "casting off."

37. And it is this characteristic of desertion that de Solà-Morales and Massimo Cacciari find in Heidegger's later work. The architect Ignasi de Solà-Morales believes that Frampton and others have misinterpreted Heidegger's notions of estrangement, desertion, and disappearance (the *unheimlich*) and as a result produced a "phenomenologically ingenuous restoration," naively focused on a tectonics at the expense of contemporary problematics of place. *Differences: Topographies of Contemporary Architecture* (Cambridge: MIT Press, 1997), 64–65. Fredric Jameson also notes that Frampton's focus on the tactile, the tectonic, and the telluric (after Heidegger) results in the reconception of space as place by displacing the visual for the invisible presence-absence of the joint. *The Seeds of Time* (New York: Columbia University Press, 1994), 197. "Paratactic" is used to describe an arrangement of terms or components without a presupposition of coordination or subordination. Derived from the Greek παράταξις, this term describes a "placing side by side" as opposed to an arranging above and below or an explicit linking. Such an arrangement relates to the open-ended and non-hierarchical quality of camping and its unique mode of relating to and producing place. The

restorative stability sought by theorists and architects such as Christian Norberg-Schulz and Kenneth Frampton is precluded by this presence of breaking. Interpreting place as a singular ontology, critics such as Norberg-Schulz, characterizing space as the systematization of place, have relied on an exteriority or a feigned objectivity in their analysis of place. Frampton's critical regionalism prioritizes the tectonic over cultural and spatial characteristics. Such treatments create a hierarchical rather than a paratactic organization of qualities of place, and thus relegate minor qualities to a peripheral role in defining place.

38. Within the cyclic process of camping, siting and "breaking into space" occur simultaneously.

39. In his reading of *Being and Time*, Casey summarizes this notion: "Dasein [being-in-the-world] takes space not only so as to 'break into space' more freely. Such an *Einbruch* into space is accomplished by making room for leeway: clearing the space for diverse engagements. From such spatial latitude, Dasein comes back to place." *Fate of Place*, 258.

CHAPTER 1. ARRIVING

1. The *mattang* is one type of map within the broader set of Marshall Island navigational devices. These "stick charts" also include the *meddo* and the *rebbelith*.

2. This sixteen-foot dimension is the external, nominal length of the trailer.

3. Martin Heidegger's reading of the Greek term *peras* defines this type of boundary.

4. It should be noted that "thresholding" is a term commonly used in processes of image alteration. In particular, thresholding is the setting of a range within a gray-scale image from which to parse out a binary coding of black and white designations. This conversion from gray-scale to a binary image must simultaneously take into account and adapt to the changing attributes of foreground and background along the image edge that is being considered and analyzed. Such "adaptive thresholding" used in digital imaging serves as one analog for the introduction of this idea of thresholding as a process of concurrent arrival and departure—a process that is ultimately a negotiation of a series of thresholds, as seen in the experience of the campsite.

5. Peter Handke, *Across*, trans. Alan Sheridan (New York: Norton, 1977), 12.

6. In chapter 11, Rudolph Schindler's Kings Road house is read as a concretization of the idea of thresholding.

7. Keri Hulme, *Te Kaihau / The Windeater* (Wellington, New Zealand: Victoria University Press, 1986), 215.

8. Umberto Eco, *The Island of the Day Before*, trans. William Weaver (New York: Penguin, 1995), 129, 148.

9. These "higher-order concepts" are often "not directly observable," but are sensed through touch, hearing, smell, and taste. See William Davenport, "Marshall Islands Navigational Charts," *Imago Mundi* 15 (1960): 22.

10. This seascape is the context (although not explicitly identified) for Deleuze and Guattari's discussion of haecceity: "there is no line separating earth and sky; there is no intermediate distance, no perspective or contour, visibility is limited; and yet there is an

extraordinarily fine topology that does not rely on points or objects but on haecceities, on sets of relations (winds, undulations of snow or sand, the song of the sand or the creaking of ice, the tactile qualities of both); it is a tactile space, or rather 'haptic,' a sonorous much more than a visual space. . . . The variability, the polyvocity of directions, is an essential feature of smooth spaces of the rhizome type, and it alters their cartography." See Gilles Deleuze and Felix Guattari, *Nomadology: The War Machine*, trans. Brian Massumi (New York: Semiotext[e], 1986), 53.

11. As defined by Deleuze and Guattari, "speed . . . constitutes the absolute character of a body whose irreducible parts . . . occupy or fill a smooth space in the manner of a vortex." Ibid., 52.

12. The navigator is "in a *local absolute* . . . engendered in a series of local operations of varying orientations: desert, steppe, ice, sea." Ibid., 54.

13. Inhabitants of the Pacific are at home on the sea, understanding both atoll and canoe as vessels of living within the watery expanse.

14. Excerpt from José Emperaire, *Les nomades de la mer* (Paris: Gallimard, 1954), 225, quoted in Deleuze and Guattari, *Nomadology*, 133; italics added.

15. Developing this idea of "station as process," Deleuze notes: "The nomad knows how to wait, he has infinite patience. Immobility and speed, . . . a 'stationary process,' station as process. . . . It is thus necessary to make a distinction between speed and movement." *Nomadology*, 51–52.

16. Davenport, "Marshall Islands Navigational Charts," 22.

17. In the Marshall Islands, the western archipelago is called Ralik, which means sunset; Ratak, meaning sunrise, is the chain of islands to the east.

18. Gilles Deleuze and Felix Guattari write of "memories of a haecceity"—possibilities for a memory of a pure experience of the moment. See *A Thousand Plateaus* (Minneapolis: University of Minnesota Press, 1987), 148–49.

19. Simon Schama, *Landscape and Memory* (New York: Knopf, 1995), 6–7. Schama speaks of his own work and research for the book as an "excavation" through which he might "recover the veins of myth and memory" beneath the surface of cultural convention (14).

20. Schama also makes the case that "cultural habits of humanity have always made room for the sacredness of nature." Ibid., 18. In the Marshall Islands, the sea frames all notions of natural consecration and mythos.

21. It should also be noted that "whorls" refer to the name of a navigational indicator as well as to the name of a ghost that was transformed into a current of this name; and the "northwest current" is the name of a current that will carry a canoe out of the region of the Marshall Islands.

22. The *Oxford English Dictionary* entry for "arriving" registers these intermediary, and at times episodic, conditions of arrival: ". . . reaching the shore, landing; arrival: act of coming to shore, landing in a country, disembarkation; a landing-place; the act of coming to the end of a journey or to some definite place; a cargo to be delivered when the ship arrives; the coming to a state of mind or stage of development; one that arrives or has arrived." 2nd ed., prepared by J. A. Simpson and E.S.C. Weiner (New York: Oxford University Press, 1989).

23. With the sublime reflectivity of the polished Airstream and the phenomenal trans-

parency of the navigational map, it could be argued that each of the constructions (though radically different in their origins and procedural relationships) serves as a filter for the context and environment.

24. Michel Serres writes, "one must seize the gesture [found in transformations, wanderings, crisscrossings] as the relation is in progress and prolong it. There is neither beginning nor end; there is a sort of vector. That's it—I think vectorially. Vector: vehicle, sense, direction, the trajectory of time, the index of movement or of transformation. Thus, each gesture is different, obviously." See Serres with Bruno Latour, *Conversations on Science, Culture, and Time,* trans. Roxanne Lapidus (Ann Arbor: University of Michigan Press, 1995), 104.

25. This radicalized empiricism, espoused by Serres, parallels Deleuze's "experimentation in contact with the real." Deleuze and Guattari's rhizomatic discipline similarly proposes the externalized forces of a nomadic thought.

26. In this case, the navigator is the Marshallese pilot, and the tourist-traveler is the Airstream owner.

27. Deleuze and Guattari, *Nomadology,* 53.

28. For Deleuze, this change is brought about by the opposition of nomadic and state forces, which are manifested in smooth and striated spaces respectively. While Deleuzian thought is inextricably political, Serresian epistemology maintains a dialogue between myth and science—for the most part eschewing classifications of doctrinal politics. The emphasis and thus focus of this discussion remains within the discourse and methodology proposed by Serres but does seek to account for the relations both implicit and overt between the two philosophers, Serres and Deleuze. To avoid a generalized political overlay, the "politics of camp" will be addressed from within the specific context of each case study. Throughout their careers, the two philosophers demonstrate mutual respect for each other's work and often cite each other in their writings. In one case, Serres describes Deleuze as "an excellent example of the dynamic movement of free and inventive thinking." *Conversations on Science, Culture, and Time,* 39.

29. A third construction that serves as a hinge between the Airstream and the *mattang* is the French author Raymond Roussel's "Maison Roulante [also written "roulotte"]." Built in the early 1920s and displayed at the 1925 Salon de l'auto in Paris, the "house on wheels" served as a mobile writing studio that assuaged Roussel's acquired phobia of luggage and allowed the author to travel virtually in place. The sumptuous interior spaces of the thirty-foot-long caravan created a hermetic volume where Roussel, with blinds drawn, could read without interruption. Mark Ford notes: "He spent his time . . . immersed in his daily ration of Loti and Verne, indifferent to the landscapes through which he was passing. The roulotte lessened still further the danger of details from his voyages seeping into his writings." *Raymond Roussel and the Republic of Dreams* (Ithaca, N.Y.: Cornell University Press, 2000), 171. Although Roussel traveled widely, his writing method and content remained at a distance from his actual experiences of the places visited so that his works such as *Locus Solus* (1914), *Impressions of Africa* (1910), and *New Impressions of Africa* (1932) retain a magically real quality. Accordingly, the relationship between his motorized caravan and his writing method warrants further study. It is my preliminary conclusion that his writing method influenced the construction and occupation of the caravan rather than the procedure of writing as a

function of the vehicle, although *New Impressions* with its parenthetical construction was begun before Roussel abandoned the caravan in late December 1926. See Roussel's *How I Wrote Certain of My Books*, ed. and trans. Trevor Winkfield (New York: Sun, 1977).

30. Ruth Deering's journal is a part of the collection in the Eaton Florida History Room at the Manatee County Central Library, Bradenton, Fla.; this entry is dated December 25, 1921. See also the beginning of chapter 6 and chapter 7.

31. An advertisement (ca. 1960) for Airstream's thirty-foot Liner travel trailer model includes the following: "The Airstream Liner is your true 'home away from home'—designed for glorious living and traveling comfort. It has the most amazing interior you've ever seen . . . a big, big living room, worlds of closet and drawer space. . . . You have the last word in livability. Wherever you stay, you will enjoy living in it. . . . Wherever you go, you will enjoy taking it with you."

32. That is to say, we did not rip out air-conditioning and throw open our windows so that we "could swelter, shiver, and struggle to hear [ourselves] above the roar of the city"; but in some cases we did "unwind by throwing paint against the walls and drilling holes into them," as Simon Sadler describes Feuerstein's activities within his "impractical flats." *The Situationist City* (Cambridge: MIT Press, 1999), 7–8.

33. In his many essays, architect and theorist Vittorio Gregotti has developed this idea, which has also been used in conjunction with the work of architect Mario Botta. The transformation of the domestic space into a construction site is not unlike a scaled-down version of Christopher Alexander's formulation of the "builder's yard." As a social institution, this "yard" decentralizes building knowledge; and as a domestic and communal component, it serves as a laboratory for construction work developed and carried out from within the community itself. Alexander refers to the "builder's yard" as the "nucleus of construction activity . . . a physical anchor point, a source of information, tools, equipment, materials, and guidance." *The Production of Houses* (New York: Oxford University Press, 1985), 94–95. The "organic relation" of the yard to its built context follows Alexander's emphasis on process rather than product and the understanding of "the building system in terms of actions that are needed to produce a building (and not in terms of its physical components" (222).

34. Viewed another way, roofs await floors and walls.

35. In *Home Territories: Media, Mobility and Identity* (New York: Routledge, 2000), David Morley has discussed the relation between the rhetorical territories of home and media, mobility, and cultural identity.

36. This rhetorical question adapts de Solà-Morales's reading of Nietzsche's understanding of the "aesthetic" in contemporary society. Rather than located in and limited to a particular place, the components of life and culture are experienced paratactically—that is to say, "side by side" without a relative or hierarchical positioning through specific places. De Solà-Morales argues that the displaced and peripheral position of aesthetic experience in contemporary culture results in a "paradigmatic value of the marginal" that forms one version of the "weak construction of the true or the real"—a construction related to his essay's central paradox that is also its title, "Weak Architecture." In *Differences: Topographies of Contemporary Architecture* (Cambridge: MIT Press, 1997), 60. For Nietzsche, this contemporary

crisis of finding a "grounding without ground," a weak architecture, occurs between the current "agitated ephemeral existence and the slow-breathing repose of metaphysical ages." *Human, All Too Human,* 2nd ed., ed. and trans. R. J. Hollingdale (New York: Cambridge, 1996), 24. Different ideas and cultures that can now be experienced in proximity without "localized domination" yield for the philosopher, as well as for de Solà-Morales's architect, an "enhanced aesthetic sensibility." But for Nietzsche, the problem remains: "A completely modern man who wants, for example, to build himself a house has at the same time the feeling he is proposing to immure himself alive in a mausoleum." Ibid., 24. It might be said that Henry David Thoreau, writing in 1854, presaged Nietzsche's concept of home: "We have built for this world a family mansion, and for the next a family tomb. The best works of art are the expression of man's struggle to free himself from this condition." *Walden* (Princeton: Princeton University Press, 2004), 41.

37. See Sigfried Giedion's *Space, Time and Architecture: The Growth of a New Tradition* (Cambridge: Harvard University Press, 1967). In my reformulation, the vernacular is the plurality of architectures in Giedion's title.

38. Edward Casey, "Smooth Spaces and Rough-Edged Places: The Hidden History of Place," *Review of Metaphysics* 51, no. 2 (1997): 268. Looking at this confusion of space and place and also operating from the premise that place is fundamentally different from space, Casey defines a distinction of place and space that is "not derivative but generative." Place is differentiated from space, and at the same time place has the potential to generate space. Casey summarizes this problem in the title of his recent book *The Fate of Place* (Berkeley and Los Angeles: University of California Press, 1997). In one sense, place's destiny has varied historically with philosophical and cultural changes, with its nadir in a fateful assimilation by space in seventeenth-century Newtonian science. In another respect, "fate" points toward place's renewed role as an event.

39. See Ignasi de Solà-Morales, "Place: Permanence or Production," in *Differences: Topographies of Contemporary Architecture* (Cambridge: MIT Press, 1997), 104.

40. Within recent work such as Bernard Tschumi's "architecture of the event" and Paul Virilio's "landscape of events," Tschumi's work seeks to address the problem of relegating event and program to chronically subordinate roles in relation to architectural space. For Tschumi, an architectural construction is not a "passive object of contemplation" but instead must be viewed as a "place that confronts spaces and actions." *Architecture and Disjunction* (Cambridge: MIT Press, 1994), 141. Consequently, architecture becomes a discourse of events. In his introduction to Paul Virilio's work, Tschumi continues this discussion: "I have always felt, as an architect, that it was more exciting to be designing conditions for events than to be conditioning designs." Foreword to *A Landscape of Events,* by Paul Virilio, trans. Julie Rose (Cambridge: MIT Press, 2000), ix. The camp provides a context for these "conditions for events."

41. Paul Virilio, "Calling Card," in *A Landscape of Events,* xi. In this way, temporality makes manifest a constructed landscape through the presencing of its particular sequence of events.

42. This problematic is particularly evident in the work of Robert Kronenburg, whose

seminal work is *Houses in Motion: The Genesis, History and Development of the Portable Building* (2002). Responses to the overall problem of mobility, ephemerality, and temporality in architecture range from an embracing of highly technological solutions to a nostalgic return to a simulacrum of stability and homogeneity. Within high technology, fabricated and highly machined details occur fleetingly inside the evacuated "non-places" of Marc Augé; and the nostalgic homecoming yields New Urbanist planned communities that sometimes adopt the form of detailing without attention to materiality and technique. And in terms of regionalism, Alan Colquhoun identifies what he terms the "core of the problem": "What is the relation between cultural patterns and technologies?" From "The Concept of Regionalism," in *Postcolonial Spaces,* ed. G. B. Nalbantoglu and C. T. Wong (New York: Princeton Architectural Press, 1997), 22. Is this relationship evolutionary or juxtaposed, or is there a dialogic middle ground in which we can engage their differences? See note 46.

43. Anthony Vidler outlines this problem of the mobile and the fixed in his discussion of John Hejduk's "vagabond architecture" that explores "a new type of space, that of the nomad, as it intersects with the more static space of established realms." See *The Architectural Uncanny* (Cambridge: MIT Press, 1992), 214. Among the many books and diary constructions of John Hejduk are *Riga* (Berlin: Aedes, 1988) and *Vladivostok* (New York: Rizzoli, 1989). In each of these works, Hejduk sounds the depths of particular places (in this case, the cities identified in the titles), in both their actual and imagined manifestations. As noted by Bernhard Schneider in *Riga*, the result of his brigade of "vagabond architecture" is "that the view from afar, from a foreigner, the distant perspective is laden with imagination which can observe many things sharper and more clearly than one familiar with the site for many years" (n.p.). Hejduk writes in *Riga* that "place actual" is transformed into "place imagined" through the "particular atmospheres and sounds" of the sites and the impregnation of his "soul with the spirit of place" (n.p.). This introduction of a characteristically heterogeneous mobile space within the normalized space of the city also summarizes the problem of ephemerality in relation to place.

44. Migrants and refugees might also be added to this group.

45. See Deborah Hauptmann, "The Past Which Is: The Present That Was," in *Cities in Transition*, ed. Arie Graafland and Deborah Hauptmann (Rotterdam: 010 Publishers, 2001). This intersection of zones of mobility and fixity potentially problematizes the relation between time and architecture. Discussing the problem of "confusions of domains of space with those of the experience of time," Hauptmann summarizes what she sees as a dilemma of architects and planners—"equivocally addressing problems of heterogeneous flow with incompatible answers in terms of homogeneous spatial fixity" (361). Thus, it might be that spaces of flow have the potential for responding to this operational disconnect, especially through a connection with issues of place and time. This research seeks to address this problem by studying the spaces and places of camps. The camp(site) begins with an openness to the particularity of place and subsequently becomes a campsite with a spatial organization arising out of the local situation.

46. In the context of regionalism and the profession of architecture, Alan Colquhoun refers to this problematic compositional appropriation of vernacular motifs as a "second-order

system" that arises from an individual architect's interpretation. See Colquhoun, "The Concept of Regionalism." The question of the vernacular brings up a related question about the role of the concept of regionalism in the production of architecture within the postmodern world. Regionalist doctrine has typically relied on the stable, public meaning of a particular place. The current dilemma of redefining regionalism arises from the "unstable difference" characteristic of contemporary places and late capitalist production. Colquhoun identifies the problem in a shift from differences between regions to differences within regions—a polyvalent condition that doctrines such as critical regionalism, with its restorative stability, are not always prepared to address.

47. Potential outcomes of vernacular as building form include reductive architectures of style, uncritical typologies, and limited morphologies. This problematic consequence is less likely with inductive (though still a posteriori) approaches in which particular situations are studied initially in terms of their own specificity and later grouped, as in the work of Henry Glassie. Glassie's analysis presents a rigorous documentation of the material culture and vernacular architecture of rural building practices in the United States (particularly the Middle Atlantic region and New England), northern Europe, and Turkey. His book *Vernacular Architecture* (2000) is an expanded version of the fifth chapter of *Material Culture* (1999) and looks comparatively at regional architecture from around the world to relate the cultural history of place. The architect Steven Holl identifies Glassie as a guide in his critical assessment of architectural types and his attempt to develop a typology suggesting "abstract context" and "unconscious logic." *Rural and Urban House Types in North America* (New York: Pamphlet Architecture, 1982), 6. Holl writes, "Glassie illuminates the dominance of geometrical ideas in the silent artifacts of indigenous rural houses the way a composer/analyst might discover the fundamental dotted dance rhythm or the structure of melodies with imperfect cadence in a folk song" (6).

48. Ignasi de Solà-Morales uses this terminology in his critique of critical regionalist models. *Differences*, 64.

49. Ibid., 65. Because of the prominence of place and methodology in this work, a treatment of "tectonic culture" remains for the most part implied in the study of detail and the way camps and their constructions are actually made. The notion of the tectonic as outlined by Kenneth Frampton in works such as *Studies in Tectonic Culture: The Poetics of Construction in Nineteenth and Twentieth Century Architecture* (1996) passes into this work through discussions of philosophers and architects such as Martin Heidegger and Gottfried Semper. See also Carles Vallhonrat, "Tectonics Considered: Between the Presence and the Absence of Artifice," *Perspecta* (1988): 122–35.

50. For Jackson's specific treatment of the vernacular's incorporation of foreign components, attributes, materials, and techniques, refer to his essay "Vernacular," in *Discovering the Vernacular Landscape* (New Haven: Yale University Press, 1984).

51. bell hooks, *Yearning: Race, Gender, and Cultural Politics* (Boston: South End Press, 1990), 148.

52. Cast as an architectural question, this issue lies between such exercises as Gordon Matta-Clark's and Rachel Whiteread's transformations of initially habitable structures into

an absence through which can be read the tactile presence of memory and traces of an occupation and acts of making. See, for example, Matta-Clark's *Splitting* (1974) and Whiteread's *House* (1993). At the other end are Krzsztof Wodiczko's projects that attempt the presencing of an absent social commentary and the production of a habitable yet mobile domestic space. An aspect pertinent to the topic of this book and its discussion of home is the proximity of strangers, familiar and unfamiliar objects, and even memories of incongruous places in Wodiczko's projections and homeless carts.

53. In an intermediate zone is the work of architects like Peter Zumthor who seek an adaptation of one home and its historical and personal memories with a newly materialized dwelling. Carried out through details and joints at the scale of the hand and the domestic volume, this synthesis of multiple homes creates a new home-place while maintaining a dialogue of invention with past and future. In the Gugalun House in Versam, Switzerland (1990–94), Zumthor believes that the project's addition to a 1706 structure forms a "new whole" and is an "absorption of new and old." Zumthor also speaks of knitting the new and old structures together, as in the tradition of Swiss vernacular log homes, and maintaining the succession of spaces also found in the local architecture.

54. One possible solution to this problem lies in an architectural understanding of the relationship among place, time, and home within zones of semipermanence.

55. One component of this hinge links architecture to the humanities.

56. Jay Fellows uses these terms to frame the work and narrative methods of John Ruskin.

CHAPTER 2. SITING CAMP

1. Susan Sontag, "Notes on Camp," in *Against Interpretation and Other Essays* (New York: Dell, 1964).

2. Occupying and moving within places combine to form a method of *practicing place.*

3. Charles Hallock, ed., *Camp Life in Florida* (New York: Forest and Stream Publishing Co., 1876).

4. See photograph of Charles Seaver in his steamboat-barge on p. 53.

5. James A. Henshall, *Camping and cruising in Florida: An account of two winters passed in cruising around the coasts of Florida as viewed from the standpoint of an angler, a sportsman, a yachtsman, a naturalist and a physician* (Cincinnati: R. Clarke and Co., 1888).

6. See Howard Lawrence Preston's *Dirt Roads to Dixie: Accessibility and Modernization in the South, 1885–1935* (1991). For a collection of Meyer's photographs (in the Alma Clyde Field Library of Florida History) and other images from the time period, refer to Nick Wynne's *Tin Can Tourists of Florida* (1999).

7. Elon Jessup, *The Motor Camping Book* (New York: Putnam's, 1921), 1.

8. Ibid., 9.

9. Emily Post, *By Motor to the Golden Gate* (New York: Appleton, 1917).

10. Jessup, *The Motor Camping Book,* 2, 189, 9, 58, 149.

11. See chapter 14 in Winfield A. Kimball and Maurice H. Decker, *Touring with Tent and Trailer* (New York: Whittlesey House, 1936).

12. Wally Byam, *Trailer Travel Here and Abroad: The New Way to Adventurous Living* (New York: David McKay, 1960).

13. Elizabeth Geniné and William Geniné, *Camping with the Family,* Public Affairs Pamphlet No. 388 (New York: Public Affairs Committee, 1966).

14. Dan Morris and Inez Morris, *The Weekend Camper* (New York: Bobbs-Merrill, 1973).

15. Gorham, B. W. *Camp Meeting Manual: A Practical Book for the Camp Ground* (Boston: H. V. Degen, 1854).

16. Ibid., vii.

17. Ibid., 13–17.

18. These sections in Gorham's manual include the arrival "Going to Camp Meeting," the duration of the stay "At Camp Meeting," and the departure "Returning from Camp Meeting."

19. Ibid., 130.

20. This dispersion is "governed by [the] hostile mechanized threat, the air situation, and control of the command." See "Camps and Bivouac Area," in U.S. War Department, *Staff Officers' Field Manual: Organization, Technical and Logistical Data* (FM 101–10) (Washington, D.C.: U.S. Government Printing Office, October 10, 1943), 901–6, 905.

21. "Shelters and Camps for Engineer Units," in U.S. War Department, *Engineer Field Manual: Operations of Engineer Field Units* (FM 5–6) (Washington D.C.: U.S. Government Printing Office, April 23, 1943), 176–80 (pars. 156–63), 176.

22. U.S. War Department, *Tents and Tent Pitching* (FM 20–15) (Washington, D.C.: U.S. Government Printing Office, February 24, 1945), 51, 8–9. Note that FM 20–15 was updated on January 9, 1956.

23. *Riga* (Berlin: Aedes, 1988), n.p. Hejduk also notes: "The objects/subjects present themselves to a city and its inhabitants. Some of the objects are built and remain in the city; some are built, stay for a while, then are dismantled and disappear; some are built, dismantled and moved to another city where they are reconstructed."

24. Federal Writers' Project, *Florida: A Guide to the Southernmost State* (New York: Oxford University Press, 1939). The *Guide to the Southernmost State* is divided into three parts: Florida's background in diverse categories from its natural features to its architecture, an overview of the state's principal cities, and a series of tours called "The Florida Loop."

25. As a collection of newspaper excerpts, oral history, text from signage, song lyrics, legend, geographic description, and statistics, the *Guide to the Southernmost State* resembles the plateau structure of Gilles Deleuze and Felix Guattari's *A Thousand Plateaus,* trans. Brian Massumi (Minneapolis: University of Minnesota Press, 1987).

26. Book 1 outlines the "preliminaries" with two epigraphs, a fable of "food, shelter, and clothing," and an inventory of "persons and places" portrayed as actors in a play. See James Agee and Walker Evans, *Let Us Now Praise Famous Men* (Boston: Houghton Mifflin, 1960).

27. The first part of book 2 specifies the time (July 1936) and summarizes the themes in "A Country Letter." Corresponding to "things that are made" and the processes of their making and use, the second part catalogues money, shelter, clothing, education, and work. After an intermission, a set of three "Inductions" develop an open-ended analysis of the findings in the previous section.

28. Agee and Evans, *Let Us Now Praise Famous Men*, 13.

29. John Ruskin, *St. Mark's Rest* (London: George Allen, 1904), 160. Ruskin states an additional objective to "examine the religious mind" of fifteenth-century Venice.

30. Robert Harbison, *Eccentric Spaces* (New York: Avon, 1980), 141.

31. Harbison writes of Ruskin's method: "His objects are works of art and his collection of them in a book a kind of museum, but he brings together a museum and a map, because he locates his objects in real space. . . . By giving the sense of a few things with lots of space around them Ruskin conceals the fact that he assembles a museum, but his powers of selection are making an order discriminate like a museum and not indiscriminate like a map, and what feels like a further freedom, leaving things where they live, is the occasion for a further order." Ibid.

32. The guide's text goes on to expound on the long-lasting virtues and use-value of this souvenir from the fair.

33. Richard J. Murphy, *Authentic Visitors' Guide to the World's Columbian Exposition and Chicago*, May 1 to October 30, 1893 (Chicago: Union News Company, 1893), 3. In addition to "permanent" and "indispensable" maps, the guide includes the text of President Harrison's proclamation, the program of the dedicatory exercises, the Exposition's chronological history, and a description of the Exposition site with its connections to transportation as well as its buildings and grounds.

34. The guide's condensed format relies on a unique "system of classification" that "has been arranged exclusively for the Authentic Visitors' Guide, and is copyrighted." Ibid., 3.

35. By pairing the officially sanctioned group numbers of exhibits (as located within Exposition structures) with the *Authentic Guide*'s page numbers, the system indexes the contents of each of the guide's pages dedicated to a particular building and department, assigned letters *A* through *N*.

36. The *Official Guide* also includes numerous illustrations adapted from "original drawings" and incorporates the "official map" of the grounds dated June 15, 1893. John J. Flinn, *Official Guide to the World's Columbian Exposition*, Souvenir Edition (Chicago: Columbian Guide Company, 1893).

37. Ibid., 5.

38. The *Official Guide* also serves as a manual to prepare for the fair with a section titled "Ten Suggestions for Visitors"—primarily a checklist of preparations, expectations, and costs for the visitor to Chicago and to the fair, with additional advice to consult with the fair's Bureau of Public Comfort upon admission.

39. M. P. Handy, ed., *Official Catalogue of Exhibits: World's Columbian Exhibition* (Chicago: W. B. Conkey, Publishers to the Exhibition, 1893). The *Official Catalogue*'s breadth of documentation contrasts with the *Authentic Guide*'s conciseness and the *Official Guide*'s lavishly illustrated documentation.

40. The introduction of each departmental section includes a frontispiece with an oblique perspectival view of the building rendered by A. Zeese and Company of Chicago.

41. The *Official Catalogue* held by the University of Florida Library includes a signature and inscription on the first page: "C. L. Willoughby . . . Corps of Guides . . . Jackson Park, Chicago . . . 1893." Willoughby has also added to the "Ground Plan of the Palace of Fine Arts"

handwritten, pencil notations that identify the spaces assigned to particular countries (see page 8 in Fine Arts section [K] of the *Official Catalogue*). The "Corps of Guides" identifies Willoughby as a member of Queen Victoria's Corps, also known as the "Frontier Force," established in 1846 as an Indian army regiment on the northwest frontier of India during British rule. Interestingly, Willoughby purchased in 1890 Plymouth Beach's Columbus Pavilion, which he enlarged to include not only room for dancing and dining but also accommodations for visitors. In 1893, he built a series of beach cottages to expand guest lodging. It is also known that Willoughby was sole proprietor around 1901 of the Cyclorama of the Battle of Gettysburg on Tremont Street in Boston.

42. To give a sense of scale, the Machinery Hall measured 1,400 by 600 feet. An interesting comparison can be made between these plans and those of the military tent camp found in the *Staff Officers' Field Manual* (FM 101–10).

43. See note 42 in chapter 3.

44. Here, the guide-manual-scrapbook resembles the diary, specifically Ruskin's *Brantwood Diaries,* in which he notes that his work will be "nothing but process." See also note 63 in chapter 5. I have expanded on these ideas about the scrapbook's relation to time and place in "Scrapbook (1923)," *Thresholds* 31 (2006): 90–101.

45. Tin Can Tourists of the World, 1920–82, unpublished documents, Florida State Archives, Tallahassee, Collection Number M93–2, boxes 2 and 3.

46. John Ruskin's self-described "caravannish manner" can also be discussed in terms of Kurt Schwitters's work with collages, reliefs, and the *Hannover Merzbau,* which was a project combining collage, sculpture, and architecture carried out between 1923 and 1937.

47. See the reference to the series in the entry for "scrapbook" in the *Oxford English Dictionary,* 2nd ed., 20 vols., prepared by J. A. Simpson and E.S.C. Weiner (New York: Oxford University Press, 1989). In its programming, the series relates to the methods used by the Federal Writers' Project participants, James Agee, and Ruskin himself. In his differentiation of modern maps and those driven more by myth, ritual, and itinerary, Michel de Certeau describes the fifteenth-century Aztec map of the Totomihuacas exodus as a "history book" rather than a "geographical map." The book portrays the log of the journey and the map of the route. Opposing the totalizing aspect of contemporary maps, de Certeau emphasizes the spatial stories inherent in the itineraries and tours that articulate the "arts of actions" and the citation of "stories of places." *The Practice of Everyday Life,* trans. Steven Rendell. (Berkeley and Los Angeles: University of California Press, 1988), 120. In this formulation, de Certeau's history book and the scrapbook both serve as a reweaving of narrative fragments through an interlaced tour of the subject, camps.

CHAPTER 3. CLEARING CAMP

1. In its use as a field, camp has been derived from the Italian, Spanish, and Portuguese term *campo.* See the entries for "camp" and "campos" in the *Oxford English Dictionary,* 2nd ed., 20 vols., prepared by J. A. Simpson and E.S.C. Weiner (New York: Oxford University Press, 1989).

2. Another more contemporary example is the Oregon Institute of Marine Biology, which

was built over two earlier campsites and field stations, one for the Corps of Engineers in the first decade of the twentieth century and the second iteration as an outpost for the Civilian Conservation Corps in the 1930s. Before the renovation in the 1990s, the institute's buildings were comprised exclusively of recycled material from these earlier constructions. See Charles Linn, "Field Station," *Architectural Record* (November 1992): 74–79. The new design includes a central green space called the "hearth," which is used for informal gatherings and volleyball and basketball games.

3. Frank Lloyd Wright's contemporaneous work with camps in Arizona might be said to have influenced his campus design for Florida Southern College, its organic and externalized spaces tied in many ways to the experience of camping.

4. The University of Florida's campus serves as a good example of the collision of university campus and military camp. See the discussion in the preface.

5. *Caeruleus campus* is literally a "dark blue level place," and *campus latus aquarium* is a "broad watery place."

6. The surface of the sea is emblematic of Gilles Deleuze's "smooth space."

7. Lillie Bernard Douglass, *Cape Town to Cairo* (Caldwell, Idaho: Caxton, 1964).

8. Gertrude Bell, *The Letters of Gertrude Bell* (New York: Penguin, 1987), 276. This particular letter to her family is dated February 17, 1911. Bell continues, ". . . It is excessively bewildering to be deprived of the use of one's eyes in this way."

9. In 1915, soon after Bell's experiences in the desert, Mark Sykes noted that "space, distance, infinite and immeasurable, is the keynote of the Assyrian landscape." See Mark Sykes, *The Caliph's Last Heritage: A Short History of the Turkish Empire* (London: Macmillan, 1915), 436.

10. Bell, *Letters*, 274. Bell attempts to see the siting of these camps in the midst of the subtle landscape as the Bedouin does: "I looked out beyond him into the night and saw the desert with his eyes, no longer empty but set thicker with human associations than any city. Every line of it took on significance, every stone was like the ghost of a hearth in which the warmth of Arab life was hardly cold, though the fire might have been extinguished this hundred years." Gertrude Bell, *The Desert and the Sown* (London: Heinemann, 1907), 60.

11. Bell, *The Desert and the Sown*, 48–49.

12. The Bedouin's navigation and understanding of the landscape could also be read as a negotiation of intensities. For Paul Virilio, this condition becomes what he calls an "aesthetics of disappearance" in which one finds the possible perception of the nonrepresentable. This idea recalls the desert as an unmappable (at least by traditional methods) landscape of subtle flows and movements. The desert then becomes a series of sublime moments, which for the Bedouin is the everyday experience of moving from campsite to campsite. See Paul Virilio, *The Aesthetics of Disappearance* (New York: Semiotext[e], 1991).

13. Drawn from references available in the *Oxford English Dictionary* (2nd ed.).

14. Homer, *Iliad*, trans. A. T. Murray (Cambridge, Mass.: Loeb Classical Library, 1999), bk. 8, sec. 489.

15. Edward Ross Wharton, *Etyma Graeca: Etymological Lexicon of Classical Greek* (Chicago: Ares, 1974), 898.

16. Homer, *Iliad*, bk. 10, sec. 198.

17. Refer to section 8, "Interchange of ω and α," in Wharton, *Etyma Graeca*, 142–43. *Chora* expresses both space and place in Plato's *Timaeus*. The connection between camp and the Greek term *chora* occurs in the possibility of the linguistic interchange between o and a (ω and α).

18. Chambers, *Cycl. Supp.*, *Camp* (1753), excerpted in the *Oxford English Dictionary*, 2: 810.

19. The date January 26, 1972, is known in Australia both as "Invasion Day" and "Australia Day."

20. See Gregory Cowan's "Nomadology in Architecture: Ephemerality, Movement, and Collaboration," University of Adelaide, 2002, http://gregory.cowan.com/nomad.

21. In a more recent example, the Camp for Oppositional Architecture has hosted a conference titled "Theorizing Architectural Resistance" in Utrecht, the Netherlands (November 10–11, 2006).

22. Summarizing this idea, Somol's essay title "The Camp of the New" (*ANY*, no. 9 [1994]: 50–55) blurs the distinction previously made by architect Colin Rowe, who divided modernist architectural response between the camp of success and the camp of the true believer.

23. Susan Sontag, "Notes on Camp," in *Against Interpretation and Other Essays* (New York: Dell, 1964), 276.

24. Ibid., 281.

25. Ibid., 286.

26. Deborah Berke, "Thoughts on the Everyday," in *Architecture of the Everyday*, ed. Steven Harris and Deborah Berke (New York: Princeton Architectural Press, 1997), 226.

27. *Oxford English Dictionary* (2nd ed.), 2: 811.

28. Mark Booth, *Camp* (London: Quartet Books, 1983), 33. We might also include method-acting, theatricality, and exaggeration to the broader meaning of *se camper*.

29. Christopher Isherwood, *The World in the Evening* (New York: Random House, 1954), 106. Isherwood continues: "Once you've done that, you'll find yourself wanting to use the word whenever you discuss aesthetics or philosophy, or just about anything. I can never understand how critics manage to do without it."

30. Lao Tze, *Tao Te Ching*, trans. D. C. Lau (New York: Penguin, 1986), VIII, lines 20–21, p. 64. I have extended the excerpt to include the line "In a home . . ." not only for its usefulness to the subject but also with a recognition of Tze's note that there is a continuity in "sense" and "rhyme" between the lines.

31. Deleuze and Guattari take up this notion of an "aphoristic style" within a multiplicity of externalized forces—the aphorism becomes the plateau. *A Thousand Plateaus*, trans. Brian Massumi (Minneapolis: University of Minnesota Press, 1987).

32. The nomadic space with its "exteriority of thought" is contrasted with the interiorized sovereignty of the State. Ibid., 45.

33. The epigraphs throughout the book provide a preliminary outline of camp's aphoristic practice.

34. Ignasi de Solà-Morales, *Differences: Topographies of Contemporary Architecture* (Cambridge: MIT Press, 1997), 59. The Italian term for "weak thought" used by Vattimo is *pensiero debole*. In translation, Vattimo's work includes *The End of Modernity* (1988) and the recently

published *After Christianity* (2002). De Solà-Morales writes: "The proposals of contemporary art are to be constructed not on the basis of any immovable reference, but under the obligation to posit for every step both its goal and its grounding" (*Differences*, 59).

35. A reading of Martin Heidegger's version of "weak thought" can be made from the term developed late in his career—*Andenken*, a "pure residuum" or recollection—that accounts for modernist impulses but remains apart from the idealism granted to the movement's goals of progress. For a discussion of this idea, see Michael Hays's introduction to de Solà-Morales's "Weak Architecture," anthologized in *Architecture Theory since 1968* (2000). Gianni Vattimo's interpretation of Heideggerian ontology as a "weak ontology" in which "the occurrence of Being is . . . an unnoticed and marginal background event" can be found in *The End of Modernity* (Baltimore: Johns Hopkins University Press, 1988), 86. See particularly pages 85–87 of the chapter "Ornament/Monument."

36. Summarizing this idea, de Solà-Morales in effect outlines the purpose and method for studying a camp architecture: "I propose it as a diagonal cut, slanting, not exactly a generational section but as an attempt to detect in apparently diverse situations a constant that seems to me to uniquely illuminate the present juncture." *Differences*, 58.

37. De Solà-Morales invokes this Nietzschean aphorism "grounding without ground."

38. Gottfried Semper, *The Four Elements of Architecture and Other Writings*, trans. Harry Francis Mallgrave and Wolfgang Herrmann (New York: Cambridge University Press, 1989), 254.

39. Expanding on his treatment of the four elements as "technical operations," Semper actually seeks to prove the thematic, as opposed to the mimetic, role of weaving that appears in both the nomadic tent and the carpet wall." See Harry Francis Mallgrave, introduction to *The Four Elements of Architecture and Other Writings*, by Gottfried Semper (New York: Cambridge University Press, 1989), 24.

40. Semper, *The Four Elements of Architecture*, 112.

41. Ibid.

42. Semper's Caribbean hut, exhibited at the Crystal Palace Exhibition of 1851, manifests his four elements of hearth, earthwork, framework-roof, and enclosing membrane in opposition to Laugier's primitive hut of 1753. Although not specifically related to the idea of modern mobility except in its transposition from one culture to another, the debate and interest sparked by this entry to the Crystal Palace exhibit parallels the later demonstration of Gypsy caravans at the World's Columbian Exposition in 1893. Also portrayed ethnographically, the display of the Gypsy caravan, coinciding with the earliest gas-powered vehicles, marks the beginnings of North America's autocamping obsession.

43. See Aaron Betsky, *Building Sex: Men, Women, Architecture and the Construction of Sexuality* (New York: William Morrow, 1995).

44. Ibid., 10.

45. Ibid., 9.

46. Ibid. Betsky continues: ". . . shading off into nature. You would have a hard time saying where this human realm begins and ends, who controls it, or what its most important seats are." For Betsky and Semper, form is not determined by space, but instead by the

operations of making and weaving that occur in and around the physical reality and idea of the camp.

47. Margaret White, *Ladies Home Journal*, June 1966, 76–77, quoted in Lynn Spigel, "Media Homes: Then and Now," *International Journal of Cultural Studies* 4, no. 4 (2001): 385–411, 392. Spigel notes that the accompanying image shows the family's indoor campout arranged around the father's miniature portable television.

48. B. W. Gorham, *Camp Meeting Manual: A Practical Book for the Camp Ground* (Boston: H. V. Degen, 1854), 14–15. Before this formalization of the camp meeting, John Wesley held what has been called the first Methodist prayer service on the American continent, in 1736 on Tybee Island near Savannah, Georgia. See the preface for a brief discussion of this event.

49. Hebraic analogies made in Methodist writings of the time compare the camp meeting to the Feast of Tabernacles, an event that today reminds Jewish people of their ancestors' living in tents or temporary huts (both of which are identified by the term *Sukkah*) in the desert. The powerful symbolism of the *Sukkah* includes the understanding of this *skhakh* as representing the clouds that protected the Israelites from the desert sun. The ritual of the *Sukkah* emphasizes how one dwells within it or otherwise occupies the structure rather than how it is constructed. In many ways, the *Sukkah* reflects the paradoxes of temporary and permanent conditions: the construction itself combines fragility and strength, and the *skhakh* combines the need for protection and openness. See also Ellen Weiss, *City in the Woods: The Life and Design of an American Camp Meeting on Martha's Vineyard* (New York: Oxford University Press, 1987), 8.

50. Sidney P. Johnston, "'No Palaces among Us': Cassadaga's Historic Architecture, 1895–1945," in *Cassadaga: The South's Oldest Spiritualist Community*, ed. John J. Guthrie Jr., Phillip Charles Lucas, and Gary Monroe (Gainesville: University Press of Florida, 2000), 96, 99. According to Johnston, the town plan was "predicated on Spiritualist ideology," and the interiors of the predominantly Victorian-style homes included rooms adapted and set aside for readings, séances, and healings (97). For the complete listing of events, attractions, and lodging available at early Cassadaga camp meetings, refer to the series of programs for the annual conventions published by the Southern Cassadaga Spiritualist Camp Meeting Association between 1912 and 1918 (in the University of Florida's Florida Heritage Collection). See also John J. Guthrie Jr., "Seeking the Sweet Spirit of Harmony: Establishing a Spiritualist Community at Cassadaga, Florida, 18933–1933," *Florida Historical Quarterly* 77, no.1 (Summer 1998): 1–38.

51. Johnston, "No Palaces among Us," 106. A similar system is found in the early stages of the formation of Braden Castle Park and will be discussed in chapter 7.

52. The original campgrounds are administered by the Southern Cassadaga Spiritualist Camp Meeting Association (SCSCMA), which has attempted to distance and differentiate itself from the wider-ranging belief systems of the practitioners outside the camp's original grounds.

53. John Brinckerhoff Jackson, "The Sacred Grove in America," in *The Necessity for Ruins* (Amherst: University of Massachusetts Press, 1980), 82, 84.

54. Ibid.

55. Latrobe quoted in Talbot Hamlin, *Benjamin Henry Latrobe.* (New York: Oxford University Press, 1955), 319–32.

56. J. B. Jackson borrows "undifferentiated" from Mircea Eliade.

57. The acronym that makes up the title *ANY* stands for "Architecture New York." Issue 1, titled "Seaside and the Real World: A Debate on Urbanism," was published in July–August 1993. In 1990, *Time* magazine referred to the town as a "down-home utopia" and "the most astounding design achievement of its era." In 1995, *Newsweek* called the town "probably the most influential resort community since Versailles."

58. See Drexel Turner, "Camptown Vitruvius: A Guide to the Gradual Understanding of Seaside, Florida," in *Urban Forms, Suburban Dreams,* ed. Malcolm Quantrill and Bruce Webb, 103–20 (College Station: Texas A&M University Press, 1993). In spite of the town's architectural fabric regulated heavily by its urban and architectural codes, Turner has expanded on this connection between Seaside and camps. Turner's title also alludes to Leon Krier's invocation of Vitruvius and neoclassicism in the town plan.

59. Malcolm Quantrill and Bruce Webb, preface to *Suburban Dreams,* ed. Quantrill and Webb, xiv.

60. With its first sessions held from February 10 to March 9, 1885, this Chautauqua community was known as the "Chautauqua that began under a tree" because the first organizers met and camped under a large live oak tree to discuss and plan the future development. See Federal Writers' Project, *Florida: A Guide to the Southernmost State* (New York: Oxford University Press, 1939), 446.

61. Chipley's subsequent decision to run the railroad next to Round Lake eventually factored into the choice of the railroad stop as the site for the Chautauquan community. The railroad access not only allowed for participants to arrive by railcar but also gave access to Chicago and New York via rail for educational and entertainment excursions. See Thomas E. Low, "The Chautauquans and the Progressives in Florida," edited by Beth Dunlap, *Journal of Decorative and Propaganda Arts, 1875–1945,* no. 23 (1998): 310–11.

62. Today, the entrance gallery is the only remaining structure of the auditorium building.

63. Turner, "Camptown Vitruvius," 114.

64. Wright wrote of the camp's components as "ephemera": "a camp we shall call it. A human inhabitant of unmitigated wilderness of quotidian change—unchangeably changing Change. For our purpose we need fifteen cabins in all. Since all will be temporary, call them ephemera. And you will soon see them like a group of gigantic butterflies, conforming gracefully to the crown of outcropping of black splintered rock gently uprising from the desert floor." *An Autobiography* (New York: Duell, Sloan and Pearce, 1943), 310.

65. See Carol Burch-Brown, *Trailers* (Charlottesville: University Press of Virginia, 1996).

66. Ibid., viii.

67. Sandra J. Stannard, "Trailers: Challenging a Tradition of Permanence and Place," *Traditional Dwellings and Settlements Review* 94 (1996): 55–70. Standard finds similar conditions in the trailer parks of Moscow, Idaho.

68. Charles Moore, "Trailers," in *Home Sweet Home: American Domestic Architecture*, ed. Charles Moore, Kathryn Smith, and Peter Becker, 49–51 (New York: Rizzoli, 1983).

69. John Brinckerhoff Jackson, "The Movable Dwelling," in *Discovering the Vernacular Landscape* (New Haven: Yale University Press, 1984), 88–101.

70. Frank Fogarty, "Trailer Parks: The Wheeled Suburbs," *Architectural Forum* 111 (July 1959): 127–31.

71. Donald Olen Cowgill, *Mobile Homes: A Study of Trailer Life* (Philadelphia: University of Pennsylvania Press, 1941).

72. Alfred A. Ring, "The Mobile Home," *Urban Land* 25, no. 7 (July–August 1966): 1–6.

73. Rudolph writes: "my thought is that the trailer industry is here to stay and that you can't get around that. It's unfortunate that the module had not been adapted to multi-story buildings. You know I've said this is the twentieth-century brick and I really believe that." See Paul Rudolph, interview with Robert Bruegmann, February 28, 1986, in "Chicago Architects Oral History Project," Ernest R. Graham Study Center for Architectural Drawings, Department of Architecture, Art Institute of Chicago, 34. Text online at: http://www.artic.edu/aic/collections/dept_architecture/rudolph.pdf.

74. Rudolph further explicates these ideas: "I'm a great believer that the vernacular architecture quite often solves problems much better than architects do. Incidentally, the trailer you see is the true vernacular of architecture in the United States whether we like it or not. One reason I'm so fascinated by it is that I've been to many trailer courts and seen what people do architecturally to what they have. I find it absolutely fascinating." Ibid., 36.

75. Rudolph leads up to this conclusion with the following: "even with the modules or the trailers, I think only about twenty-eight percent of them actually get moved from a site because making the foundations and hooking them up with water is too great an expense." Ibid., 34.

76. Peter Cook, "Plug-In," in *Archigram*, ed. Cook (New York: Princeton Architectural Press, 1999), 39.

77. Ibid., 21.

78. Greene's projects are roughly contemporaneous with the work of Marx, who published *The Machine in the Garden* in 1964.

79. David Greene, "Gardener's Notebook," in *Archigram*, ed. Peter Cook (New York: Princeton Architectural Press, 1999), 110–15.

80. See also Henri Lefebvre, *The Production of Space* (Oxford: Blackwell, 1991).

81. Johan Huizinga, *Homo Ludens: A Study of the Play-Element in Culture*, trans. R.F.C. Hull (Boston: Beacon Press, 1955), 21.

82. Ibid., 8. One premise for the construction of such projects as Constant's New Babylon and Archigram's "free time nodes" was that technological innovation allowed for an increase in leisure time. At the same time, the Situationists and their precursors in Cobra and the Lettrist International opposed the all-encompassing functionalism that they saw in the Bauhaus and other contemporaneous architectural movements.

83. Lefebvre defines the *fête* as "the eminent use of the city, that is, of its streets and squares, edifices and monuments." *La Fête* is "a celebration which consumes unproductively,

without other advantages but pleasure and prestige and enormous riches in money and objects." *The Production of Space,* 66.

84. Ibid., 173. For a discussion of the relationship between *habitat* and *inhabit* through a Heideggerian lens, refer to pages 76–80 in Lefebvre's *Writings on Cities* (trans. Eleonore Kofman and Elizabeth Lebas [Oxford: Blackwell, 1995]). It is also important to point out that Lefebvre's previous work addresses his premise that modern space is political or, more precisely, politicized with an emphasis on rationalist functionalism at the expense of a poetics or lyricism of living or "inhabiting."

85. These quantitative features of the camp include the size of the lot and the resale value—all economies of the real estate of the fixed house. Finally, fragments such as the "Situation Gloop" in Archigram's Living City make more direct allusion to the Situationists' concurrent discourse across the English Channel. The Situation Gloop is an area of exhibition concerned with the "happenings within spaces in-city [and] the transient throw-away objects, the passing presence of cars and people." Such situations can be the result of an individual, or they can be precipitated by the masses. Archigram and the Situationists shared the objective of transcending the banal through the freedom of play. Camps have played an important role as the physical and theoretical model for making a place of play.

86. Constant Nieuwenhuys, "New Babylon" (1974), in *Theory of the Dérive and Other Situationist Writings on the City,* ed. Libero Andreotti and Xavier Costa (Barcelona: Museu d'Art Contemporani de Barcelona, 1996).

87. Ibid.

88. Troels Andersen quoted in Mirella Bandini, "An Enormous and Unknown Chemical Reaction: The Experimental Laboratory in Alba," in *On the Passage of a Few People through a Rather Brief Moment in Time: The Situationist International 1957–1972,* ed. Elisabeth Sussman (Cambridge: MIT Press, 1989), 71.

89. The exhibition's space was draped with hundreds of feet of Pinot-Gallizio's unrolled paintings and canvases to simulate an immersion in the urban environment that the artist called the "cavern of anti-matter." See Peter Wollen, "Bitter Victory: The Art and Politics of the Situationist International," in *On the Passage of a Few People through a Rather Brief Moment in Time: The Situationist International 1957–1972,* ed. Elisabeth Sussman (Cambridge: MIT Press, 1989), 50. The complete excerpt includes more details of the construction: "free time, rather than being filled with banality . . . could be occupied in creating brightly painted autostrade, massive architectural and urbanistic constructions, fantastic palaces of synesthesia, the products of industrial poetry, and sites of magical-creative-collective festivity."

90. Ibid., 23–25.

91. Constant Nieuwenhuys, "New Urbanism," *Provo,* no. 9 (1966), English translation published by the Friends of Malatesta (Buffalo, N.Y., 1970), http://www.notbored.org/new-urbanism.html.

Chapter 4. Making Camp

1. Dante Alighieri, *De vulgari eloquentia,* ed. and trans. Steven Botterill (New York: Cam-

bridge University Press, 1996), 1.9.6, p. 21. Ironically, Dante must use Latin, which he believes is artificial in contrast to the vernacular's natural quality, to outline his proposal for the "illustrious vernacular": "Nec aliter mirum videatur quod dicimus quam percipere iuvenem exoletum quem exolescere non videmus."

2. This aspect of vernacular as process relates to Michel Serres's discussion of the organism-system-sheaf. In the "Origin of Language," Serres notes, an "organism is a system" and later asks, "What is an organism? A sheaf of times. What is a living system? A bouquet of times." *Hermes: Literature, Science, and Philosophy,* ed. Josué V. Harari and David F. Bell (Baltimore: Johns Hopkins University Press, 1982), 71, 75.

3. This concept of locality can also be found in the explicit etymology of its synonym "autochthonous"—meaning "from one (or the same) earth or ground." *Oxford English Dictionary,* 2nd ed., 20 vols., prepared by J. A. Simpson and E.S.C. Weiner (New York: Oxford University Press, 1989).

4. In this respect, Dante's vernacular is similar to a *koin'* or a *lingua franca* in its attempt to gather together linguistic and syntactic fragments from various dialects into a standard language common to a larger region.

5. Dante writes: "So this is why those who frequent any royal court always speak an illustrious vernacular; it is also why our illustrious vernacular wanders around like a homeless stranger, finding hospitality in more humble homes—because we have no court." *De vulgari eloquentia,* 1.18.3.

6. Dante finds himself in a similar situation to that of the vernacular as a result of his exile from Florence beginning in January 1302: "To me, however, the whole world is a homeland, like the sea to fish. . . . I suffer exile unjustly—and I will weight the balance of my judgment more with reason than with sentiment." Ibid., 1.6.3. Note that Dante's writing of *De vulgari eloquentia* between 1303 and 1305 immediately follows his condemnation to exile in 1302. In his wandering life, Dante becomes the troubadour from whom he borrows in researching the "illustrious vernacular."

7. J. B. Jackson discusses these transient communities: "It was not simply rural and agricultural; it was identified with mining and shipping communities, with cities and architect- or engineer-planned villages having military or political function. Finally, it used materials and techniques imported from elsewhere." *Discovering the Vernacular Landscape* (New Haven: Yale University Press, 1984), 86.

8. Michel Butor, *The Spirit of Mediterranean Places,* trans. Lydia Davis (Marlboro, Vt.: Marlboro Press, 1986), 23–24.

9. This idea of improvisation found in the vernacular is differentiated from the ad hocism of Charles Jencks and Nathan Silver in *Ad Hocism: The Case for Improvisation* (New York: Doubleday, 1972).

10. The concept of "vernacular as process" relates to jazz improvisation. See Mark Levine's *The Jazz Theory Book* (1996) and Jerry Coker's *Improvising Jazz.* Coker discusses jazz improvisation as having five elements in which the intellect (the "only completely controllable factor") understands the general framework, or improvisational system, which is then transformed and modified by the other four, more subconscious, factors—intuition, emotion,

sense of pitch, and habit. *Improvising Jazz* (New York: Simon and Schuster, 1987), 4. The intellectual "coding" is tempered by the emotive practice of playing, but at the same time the improvisation requires the establishment and understanding of a system of harmonic construction, tone, individual chords, and scales. For a treatment of jazz improvisation and its relationship to contemporary culture, see also Ted Gioia's *The Imperfect Art: Reflections on Jazz and Modern Culture* (New York: Oxford University Press, 1990). Gioia proposes an "aesthetics of imperfection" to discuss jazz as an art that privileges the "haphazard" over the "premeditated." In his essay "Hiphop Rupture" for the journal *Ctheory*, Charles Mudede takes up Gioia's claim and pursues the idea that accident and mistake are essential compositional components of hip-hop music. Mudede identifies three types of error: rupture (the moment of a song's collapse); incidental noise (in which fragmentary noises are "dropped" into the flow of a track); and "the art of wrecking records." http://www.ctheory.net/text_file.asp?pick=225.

11. See Sylvia Lavin, *Quatremère de Quincy and the Invention of Modern Language of Architecture* (Cambridge: MIT Press, 1992).

12. Edward Casey, *The Fate of Place* (Berkeley and Los Angeles: University of California Press, 1997), 302.

13. Emphasis on the building, in many historical and contemporary treatments, results in stylistic, formal, or typological analysis and an eventual and often uncritical appropriation.

14. Chapters 5 through 10 in this volume are each dedicated to the study of a particular place and its camp constructions. The sequence of their presentation moves thematically through each place, with an initial focus on the southern United States leading to chapter 9's geographic extension to the western United States. The "making" of this sequence proceeds "from camp to camp." In part 3, the final two chapters summarize and rethink the implications of camping for contemporary place-making through the connections and cross-references among the camps.

15. See the discussion in chapter 5.

16. Two objects of study, however, will require a more historical assessment. Manila Village was destroyed and must be studied wholly from the traces present in maps, narratives, and histories of the place. The municipal camps in chapter 6 have for the most part closed or have been modified or privatized and are studied as precursors of more permanent private camps like Braden Castle Park.

17. Paul Virilio, *Bunker Archeology* (New York: Princeton Architectural Press, 1994), 11, 13. Archaeological stratigraphy blurs distinctions of linear time so that our experiences of place are close to Bergson's "mobility in contact with the real."

18. Here "makeshift" connotes the improvisational character of making things with temporary substitutes—a kind of bricolage without always producing an assemblage. In the entry for "-ic," the *Oxford English Dictionary* (2nd ed.) notes that the Greek suffix <Greek>–46`*H*, as one of the "commonest" suffixes, serves as a "living formative" for an extensive array of descriptive terms.

19. The devices also expose what we might be "thinking" about as we camp and participate in the book's "camping space."

20. Raoul Bunschoten, "The Skin of the Earth: A Dissolution in Fifteen Parts," *AA Files*, no. 21 (Spring 1991): 57.

21. Between 1303 and 1305, Dante writes *De vulgari eloquentia*, and in 1970, Labov publishes his seminal work *The Study of Nonstandard English*.

22. This ordering is "the proportion to scale of the work's individual components taken separately, as well as their correspondence to an overall proportional scheme of symmetry." See Vitruvius, *Ten Books of Architecture*, trans. Ingrid D. Rowland, ed. Rowland and Thomas Noble Howe (Cambridge: Cambridge University Press, 2001), 24. Vitruvian *taxis* could also inform a review of the types of modular housing units that comprise many of the camps studied.

23. See Michel Serres, *The Parasite*, trans. Lawrence R. Schehr (Baltimore: Johns Hopkins University Press, 1982).

24. This series of studies seeks to "watch how things are made" as it moves from campsite to campsite. The analytic-poetic operators presented in each chapter title and throughout the analyses and proposals can be read both as descriptive terms and as exploratory terminology. The mode and manner proposed by each term outlines a working definition of camping from which the investigation into poetic occupations and modifications of place can proceed. Our itinerary takes us from taxonomic to asymptotic to parasitic to eutopic to heterotopic.

25. C. S. Peirce describes the diagram as an "icon of intelligible relations." See the discussion in chapter 5.

CHAPTER 5. CAMP(SITE)

1. Gibsonton and the sites of the Tin Can Tourists will be treated with greater detail in later chapters.

2. Guides for camping are typically framed as manuals for surviving the natural elements or familial relations and include information on preparation (food, safety, and equipment) and practice.

3. Writer Stetson Kennedy anchors the group of researchers involved in recording Florida's complex history.

4. "Caution to Tourists," in the Federal Writers' Project, *Florida: A Guide to the Southernmost State* (New York: Oxford University Press, 1939), xxii.

5. Implicit in this warning statement, and throughout the guide's many tales, is the idea that the region's interactions and inversions of the temporary and the permanent necessitate the development of an alternative working method for occupying its shifting landscape.

6. Making the case for vernacular as a negotiating procedure, this chapter also serves as a guide or manual for studying and occupying "camp space."

7. James A. Henshall, *Camping and cruising in Florida: An account of two winters passed in cruising around the coasts of Florida as viewed from the standpoint of an angler, a sportsman, a yachtsman, a naturalist and a physician* (Cincinnati: R. Clarke and Co., 1888), 41. Gainesville: University of Florida, Florida Heritage Collection, http://fulltext.fcla.edu/cgi/t/text/text-idx?c=fhp&idno=FS00000044&format=pdf.

8. Among his many works, Kennedy authored *Palmetto Country* (1942) and *The Klan Unmasked* (1954).

9. Stetson Kennedy, "Turpmtine" [*sic*], in *Palmetto Country* (Tallahassee: Florida A&M University Press, 1989), 257–68.

10. Zora Neale Hurston, "Proposed Recording Expedition into the Floridas," Library of Congress, WPA Collections, 1939, http://memory.loc.gov/cgi-bin/query/r?ammem/flwpa:@field(DOCID+essay1). Of the components in her proposal, Hurston only completes recordings at turpentine camps in Cross City and Tampa.

11. The explanatory and the exploratory are two trajectories of this chapter and this research as a whole.

12. Many more remained obscured by foliage and by their oftentimes small scale. More recently, planners have identified mobile home settlements as "stealth housing."

13. Welman A. Shrader, ed., *Florida from the Air* (Vero Beach, Fla.: Aero-Graphic Corp., 1936), 3.

14. The icons used in the legend were derived from their usage in Dan Morris and Inez Morris's *The Weekend Camper* (New York: Bobbs-Merrill, 1973).

15. The first sections of this chapter reveal and review existing spaces of camping in the region of central Florida around Tampa Bay. Subsequent intermediate sections, "Siting Camp" and "Clearing Camp," put forward practical and theoretical extensions of the identified camp spaces by demonstrating the literal and metaphorical operations that are associated with camping. The final section proposes a practice derived from Florida's camps for inventing and constructing spaces—that is, methods of researching and simultaneously occupying Florida's cultural and architectural context.

16. Wallace Stevens, "Nomad Exquisite," in *The Palm at the End of the Mind*, ed. Holly Stevens (New York: Vintage, 1990), 44.

17. Monroe founded the magazine in Chicago in 1912.

18. Stevens, *The Palm at the End of the Mind*, 401. This original line appears in a notation about the poem at the conclusion of the collection. Daryl Hine, e-mail to the author, November 26, 2002.

19. Gaston Bachelard discusses these ideas in *Water and Dreams*, trans. Edith R. Farrell (Dallas: Pegasus, 1983).

20. In *Water and Dreams*, Bachelard notes that "[d]reams come before contemplation. Before becoming conscious in sight, every landscape is an oneiric experience" (4).

21. Bachelard writes: "The reverie works in a star pattern. It returns to its center to shoot out new beams." *The Psychoanalysis of Fire*, trans. Alan C. M. Ross (London: Routledge and Kegan Paul, 1964), 14.

22. Ibid., 16.

23. James B. Finley, *Autobiography* (1853), quoted in John Brinckerhoff Jackson, "The Sacred Grove in America," in *The Necessity for Ruins* (Amherst: University of Massachusetts Press, 1980), 77–88, 86. Jackson excerpts the following: "At night, the whole scene was awfully sublime. The range of tents, the fires, reflecting light amid the branches of the towering trees; the candles and lamps illuminating the encampment, hundreds moving to and fro,

with lights or torches, like Gideon's army; the preaching, praying, singing, and shouting, all heard at once, rushing from different parts of the ground, like the sound of many waters, was enough to swallow up all the powers of contemplation." See the discussion in chapter 3.

24. Federal Writers' Project, *A Guide to the Southernmost State*, 137.

25. Ibid., 403.

26. Camping entails the problems of deciphering a place when only traces remain. Maps without legends become legendary in their ambiguity, difficulties of reading, and multiplication of possible interpretations. Reading its ashes or listening to the stories told around the campfire suggest one mode for interpreting places. Mixing fact and legend, the following preliminary stories about each place serve as inchoate travel guides to what is the shifting syntactic field of the campground. This section carries out at a smaller synoptic scale the documentary activity of the Federal Writers' Project's *Guide*, which presents information gleaned from a variety of sources without privileging one resource over another. In this paratactic mode, the *Guide* juxtaposes fragments of folklore with economic facts and accounts of quotidian life with descriptions of annual festivals.

27. Nick Wynne, *Tin Can Tourists of Florida 1900–1970*, Images of America series (Charleston, S.C.: Arcadia, 1999), 7.

28. Lori Robinson and Bill de Young, "Socialism in the Sunshine: The Roots of Ruskin, Florida," *Tampa Bay History* 4, no. 1 (Spring/Summer 1982): 5.

29. Richard Stanaback, *A History of Hernando County* (Brooksville, Fla.: Hernando County Historical Society, 1976), 91.

30. Excerpt from a history of Masaryktown, Florida, in the Federal Writers' Project, *Guide to the Southernmost State*, 391.

31. The word "legend" has its origins in the Latin word *legere*, meaning "to read."

32. The work of the architect Adolf Loos with domestic space parallels this use of interstitial space. The internal spaces of Loos's Villa Muller (1930) do not have conventionally defined thresholds between rooms. The resulting *Raumplan*, found in many of Loos's projects but only identified by his associate, works with a process of movement through spaces that can be compared to "thresholding" within the campsite, discussed in chapter 1.

33. Jay Fellows, *The Failing Distance: The Autobiographical Impulse in John Ruskin* (Baltimore: Johns Hopkins University Press, 1975), vii.

34. Jay Fellows, *Ruskin's Maze: Mastery and Madness in His Art* (Princeton: Princeton University Press, 1981), 72.

35. Robert Plunket, "Walker Evans, the Mangrove Coast, and Me," in *Walker Evans: Florida*, by Walker Evans (Los Angeles: J. Paul Getty Museum, 2000), 9–12. See the discussion of Agee's method in chapter 2.

36. Gilles Deleuze and Felix Guattari, *Nomadology: The War Machine*, trans. Brian Massumi (New York: Semiotext[e], 1986), 110–15.

37. Bonnie G. McEwan, ed., *The Spanish Missions of La Florida* (Gainesville: University Press of Florida, 1993).

38. John H. Hann and Bonnie G. McEwan, *The Apalachee Indians and Mission San Luis* (Gainesville: University Press of Florida, 1998).

39. This gathering differs from Martin Heidegger's gathering of the fourfold. The gathering found in camping practice is fragmentary and, with a chronically impending departure, always incomplete.

40. Peirce uses the example of a painting: "So in contemplating a painting, there is a moment when we lose the consciousness that it is not the thing, the distinction of the real and the copy disappears, and it is for the moment a pure dream—not any particular existence and yet not general. At that moment we are contemplating an *icon*." See Charles Sanders Peirce, *The Collected Papers,* ed. Charles Hartshorne and Paul Weiss (Cambridge: Belknap Press of Harvard University Press, 1960), 3: 211, par. 362; Peirce's emphasis.

41. See Gregory Ulmer, "Metaphoric Rocks: A Psychogeography of Tourism and Monumentality," in *The Florida Landscape: Revisited,* ed. Christoph Gerozissis (Lakeland, Fla.: Polk Museum, 1992), and "Abject Monumentality," *Lusitania* 1 (1993): 9–15.

42. Diana Knight, *Barthes and Utopia: Space, Travel, and Writing* (Oxford: Clarendon Press, 1997), 2, 12. Knight uses the phrase "utopic impulse" to describe Roland Barthes' ideas about space, travel, and writing.

43. John Ruskin, *The Works of John Ruskin,* ed. E. T. Cook and Alexander Wedderburn, Library Edition (London: George Allen, 1903–12), 25: 216. See also note 70 in chapter 2.

44. "To snare a sensibility in words, especially one that is alive and powerful, one must be tentative and nimble. The form of *jottings,* rather than an essay (with its claim to a linear, consecutive argument), seemed more appropriate for getting down something of this particular *fugitive* sensibility." See Susan Sontag, "Notes on Camp," in *Against Interpretation and Other Essays* (New York: Dell, 1964), 276–77. Emphasis has been added for comparison to Ruskin's work.

45. Ibid., 281.

46. Related to the notion of camp is Michel Foucault's usage of "imaginative heterotopia" to describe and classify the phenomenon of the colony, which for him includes the "placeless place" of the ship. See Foucault, "Other Spaces," *Lotus International,* nos. 48–49 (1986): 14.

47. In contrast to Foucault's heterotopia, Vittorio Gregotti's atopia includes both what he calls "atopical typologies" and "residual spaces." Service stations characterize the former, and junkyards the latter. See Vittorio Gregotti, "On Atopia," in *Inside Architecture,* trans. Peter Wong (Cambridge: MIT Press, 1996), 80.

48. Ibid., 81.

49. Lucy Lippard, *On the Beaten Track: Tourism, Art, and Place* (New York: New Press, 1999). See also Lucy Lippard, *The Lure of the Local: Senses of Place in a Multicentered Society* (New York: New Press, 1997).

50. Ignasi de Solà-Morales, "Place: Permanence or Production," in *Differences: Topographies of Contemporary Architecture* (Cambridge: MIT Press, 1997), 102.

51. Michel Serres, *Genesis,* trans. Geneviève James and James Nielson (Ann Arbor: University of Michigan Press, 1995), 19.

52. Codes restrict such elements as publicly visible drapery and colors of house paint.

53. See the image of the rules and the related discussion preceding chapter 1.

54. Michel Serres has referred to this type of space as a "broken threshold." See Michel

Serres, "Language and Space: From Oedipus to Zola," ed. Josué V. Harari and David F. Bell, *Hermes: Literature, Science, and Philosophy* (Baltimore: Johns Hopkins University Press, 1982), 47. The discussion will return to the term "broken threshold" in the concluding chapter.

55. W. H. Hesselman, "Ode to the TCT [Tin Can Tourists]," Collection of Florida State Archives, Tallahassee.

56. See the Land Use and Zoning section of Hillsborough County's Planning and Growth Management Office, http://was.hillsboroughcounty.org/pgm_zoning/mapkey.html.

57. Plunket, "Walker Evans, the Mangrove Coast, and Me," 41. The history of Gibsonton's zoning will be expanded and critically assessed in chapter 8.

58. In his essay "The Movable Dwelling" (in *Discovering the Vernacular Landscape*, 101), J. B. Jackson seeks to understand the "new kind of home we are all making in America." From this text, two hypothetical relations between mobility and home can be gleaned. One, the American house is a temporary construct that, though conceived initially to enclose a mobile notion of home, can gain permanence over time. Second, though most Americans are housed in semipermanent dwellings, the American concept of home is temporary, fleeting, and mobile. These hypotheses begin to outline a network of relations that exist between the two terms of the phrase "mobile home."

59. De Certeau's interpretation of Wassily Kandinsky's "metaphorical city" will be discussed in chapter 10.

60. Michel de Certeau, "Spatial Stories," *The Practice of Everyday Life*, trans. Steven Rendall (Berkeley and Los Angeles: University of California Press, 1988), 115–30.

61. The term "perimeter projects" was used by Robert Segrest in his article "The Perimeter Projects: Notes for Design" in *Assemblage* (1986) and later in "The Perimeter Projects: The Architecture of the Excluded Middle" in *Perspecta 23: Yale Architectural Journal* (1987). As will be explored in later chapters, Segrest's discussion of the carnival midway relates to this definition of camp.

62. The referenced passage occurs in the appendix of volume 35 of *The Works of John Ruskin*. The section titled "The Tour of 1841" is an appendix to Ruskin's diaries of his travels through Venice, Padua, Milan, Turin, Susa, Mont Cenis, and finally Lans-le-bourg, which (along with Chamouni) he considers as one of his "homes on earth." In an earlier passage, he identifies Rouen, Geneva, and Pisa as "centers of his life."

63. *The Works of John Ruskin*, 35: 619. It should be noted that Ruskin is composing this work in 1882 from diary fragments (1841–42) and recollections of his previous experiences. The textual context for the parenthetical statement (I will return . . .) is the following: "In the first place [city of Laon, in 1842], I had the invaluable quality of ductility. In fact, I was a mere piece of potter's clay. . . . In the second place [Laon, 1882], I had a curiously broad scope of affection, alike for little things and large. From my ants' nest in Herne Hill garden, up to Mont Blanc and Michael Angelo [*sic*], nothing came amiss to me. . . . I liked small things for being small . . .; the weak for their weakness. . . . And with this power of adaptation, I had also a sensual faculty of pleasure in sight, as far as I know unparalleled. Turner very certainly never took the delight in his own drawings that I did, else he had more uniformly drawn beautiful and sublime things, instead of, as too often, merely intellectually true ones

(I will return to this point afterwards), and certainly he would often have painted subjects for his own pleasure, instead of waiting for commissions. Ductility, comprehensiveness, sensitiveness—and associated with this third, horror of pain and disorder—leading me to wide human compassion; then fourthly, intense delight in . . . physical science" (618–19). It is very interesting to note that Ruskin is writing about a return to a particular place and that the parenthetical statement also implies a subsequent return to a particular place in the text—a revisiting of the recollection he is making. This circularity, with its spiral tendency and attendant transformative effect, is shared with the camping project.

64. Ruskin's diagram might be compared to this chapter's legend and then might also be reinterpreted:

$$\frac{(\text{site})^1}{{}^1\text{camp}}$$

65. See Fellows, *The Failing Distance.*

66. Deleuze and Guattari have developed ideas on the invention of a territoriality and the processes of reterritorialization. See *Nomadology: The War Machine,* 74, 50.

67. Ibid., 54.

68. Ibid., 54–55.

69. De Solà-Morales, "Place: Permanence or Production," 104. These conjectural foundations materialize again in the platform communities of Manila Village in Louisiana's Barataria Bay.

CHAPTER 6. MOVING IMAGES OF HOME

1. The journey between De Soto Park and Mr. Macklin's probably took at least an hour each way given the difficult road conditions of the time.

2. Entry for "municipal" in the *Oxford English Dictionary,* 2nd ed., 20 vols., prepared by J. A. Simpson and E.S.C. Weiner (New York: Oxford University Press, 1989). This quality of having narrow limits comes from the Latin term *di municipes,* which identifies gods who were worshiped only in particular localities.

3. See Walter Weyrauch, *Gypsy Law: Romani Legal Traditions and Culture* (Berkeley and Los Angeles: University of California Press, 2001). Refer particularly to Weyrauch's section titled "Parallels between Private Lawmaking and Tribal Law," in which he discusses the relationship between Gypsy rules and laws and the larger tradition of Western lawmaking practices (19–21).

4. This idea returns to the Greek origin of colony as α′ποικοι. Michel de Certeau addresses this idea of the "foreigner-at-home" in terms of language. De Certeau looks to Wittgenstein's philosophical work with linguistics: "In the accidental ways of being a foreigner away from home (like any traveler or keeper of records) Wittgenstein sees the metaphors of foreign analytical procedures inside the very language that circumscribes them. . . . This is no longer the position of professionals, supposed to be civilized men among savages; it is rather the position which consists in being a foreigner at home, a 'savage' in the midst of ordinary culture, lost in the

complexity of the common agreement and what goes without saying." *The Practice of Everyday Life*, trans. Steven Rendall (Berkeley and Los Angeles: University of California Press, 1988), 13.

5. Barbara Hunting, *Tampa Tribune*, 1960.

6. Howard Lawrence Preston, *Dirt Roads to Dixie: Accessibility and Modernization in the South, 1885–1935* (Knoxville: University of Tennessee Press, 1991), 116.

7. Allan D. Wallis, *Wheel Estate: The Rise and Decline of Mobile Homes* (New York: Oxford University Press, 1991), 40.

8. Kenneth Roberts published "The Time-Killers" on April 1, 1922, and "The Sun-Hunters" on April 15, 1922.

9. Preston, *Dirt Roads to Dixie*, 122.

10. Donald O. Cowgill, *Mobile Homes: A Study of Trailer Life* (Washington, D.C.: American Council on Public Affairs, 1941).

11. Wallis, *Wheel Estate*, 71.

12. The Tin Can Tourist falls between classification as vacationer and stranger in much the same way that John Hejduk works between familiarity and defamiliarized contexts, between the traveler and the outsider. The Tin Can Tourists' camps approach this "vagabond architecture" of the *unheimlich*, combining home with a sublime detachment. See Anthony Vidler's "Vagabond Architecture," in *The Architectural Uncanny*, by Vidler, 207–16 (Cambridge: MIT Press, 1992).

13. Mary Anne Beecher, "The Motel in Builder's Literature and Architectural Publications: An Analysis of Design," in *Roadside America: The Automobile in Design and Culture*, ed. Jan Jennings, 115–24 (Ames: Iowa State University Press, 1990), 116.

14. "Roadside Cabins for Tourists," *Architectural Record* (December 1933): 457.

15. Preston, *Dirt Roads to Dixie*, 123.

16. Ibid., 124.

17. Wallis, *Wheel Estate*, 41.

18. A visible addition to this cartoon is a penciled inscription on the woman's apron—the initials "E.H.C.W.," perhaps identifying a member of the organization.

19. Roberts, "The Sun-Hunters," 84–85.

20. For example, the Camping Tourists of America, an offshoot of the Tin Can Tourists that founded Braden Castle Park, included members with the title of "Doctor," possibly alluding to either medical doctors or academics. Also, a 1936 issue of *Trailer News* calls the trailerites "the great middle class of our people" (Kavanaugh).

21. The original documents are housed in the Florida State Archives in Tallahassee.

22. Preston, *Dirt Roads to Dixie*, 122.

23. See Nick Wynne's *Tin Can Tourists of Florida 1900–1970*, Images of America series (Charleston, S.C.: Arcadia, 1999).

24. This residence was established through habitual dwelling in a foreign place.

25. In the late 1920s and early 1930s, the season began with an annual "homecoming" celebration that continued from mid-November to Thanksgiving in Dade City, Florida. Federal Writers' Project, *Florida: A Guide to the Southernmost State* (New York: Oxford University Press, 1939), 117, 371, 537. The group celebrated another homecoming known as the "Tin Can-

ners' Homecoming" with an annual Christmas convention at the Municipal Tourist Camp in Arcadia, Florida (ibid., 370–71).

26. Ibid., 371.

27. Lucy Lippard, *On the Beaten Track: Tourism, Art, and Place* (New York: New Press, 1999), 13. Lippard expands on David Harvey's notion that tourism is about becoming rather than being.

28. The authors of the *Guide to the Southernmost State* go so far as to characterize the Tin Can Tourists' homecoming in Dade City as an "annual hegira," invoking a secularized version of Mohammed's flight from Mecca to Medina (371). Less specifically, the term describes the flight from danger, in this case, the climatic onslaught of the northern winters.

29. "Wally Byam Creed," in Bryan Burkhart and David Hunt, *Airstream: The History of the Land Yacht* (San Francisco: Chronicle, 2000), 83.

30. Two of these journeys are documented in published narratives: Lillie B. Douglass, *Cape Town to Cairo* (Caldwell, Idaho: Caxton, 1964); and McGregor W. Smith Jr., *Thank You, Marco Polo: The Story of the First Around-the-World Trailer Caravan* (Coral Gables, Fla.: Wake-Brook House, 1966), and in Wally Byam's own publication, *Trailer Travel Here and Abroad: The New Way to Adventurous Living.* The "Caravan around the World" is also documented in a 1966 video narrated by Vincent Price.

31. This movement of discovery can also be found in the early projects of the Italian Superstudio group, particularly in their vignette "A Journey from A to B" (1969), for which their caption reads: "there will be no further reason for roads or squares." This projected world would include a series of rest stops similar to Byam's "Land Yacht Harbors."

32. See Wally Byam, *Trailer Travel Here and Abroad.* A comparison can be made between these harbors and Constant's Model for a Gypsy Camp.

33. Vaïda Voivod III in an interview published in the *Algemeen Handelsblad* (Amsterdam), May 18, 1963, in Constant Nieuwenhuys, "New Bablyon" (1974), in *Theory of the Dérive and Other Situationist Writings on the City,* ed. Libero Andreotti and Xavier Costa (Barcelona: Museu d'Art Contemporani de Barcelona, 1996), 154.

34. Ibid., 156–57.

35. Ibid., 157. Peter Wollen notes that the Situationists saw the festival as a method for and a process within this idea of freedom: "[t]hey stressed the role of the creative impulse, of art as an expression of an attitude to life, dynamic and disordered like a popular festival, rather than a form of identical production." "Bitter Victory: The Art and Politics of the Situationist International," in *On the Passage of a Few People through a Rather Brief Moment in Time: The Situationist International 1957–1972,* ed. Elisabeth Sussman (Cambridge: MIT Press, 1989), 41.

36. Wally Byam's "Four Freedoms" are included in Burkhart and Hunt's *Airstream,* 42.

37. Though clearly arising from different Marxist and democratic-social political origins and drawing directly and indirectly from Huizinga, the ideas espoused by Constant and Byam for a freedom to travel and to experience are both tied to and derived from the productivity-efficiency of modern technology—Byam's inventive use and promotion of the automobile-trailer and Constant's understanding of the general automation-mechanization of work.

38. A presence actual and today virtual on the Internet; see www.tincantourist.org.

39. The Sarasota City Trailer Camp was also known in its various permutations as Sarasota Mobile Home Park, Sarasota Tourist Park, and Sarasota Municipal Tourist Camp.

40. "Mobile Home Park Believed Oldest," *Sarasota Herald-Tribune,* November 15, 1964.

41. John Nolen Papers, Department of Manuscripts and University Archives, Cornell University Library, in Michael McDonough, "Selling Sarasota: Architecture and Propaganda in a 1920s Boom Town," *Journal of Decorative and Propaganda Arts, 1875–1945,* no 23. (1998): 22.

42. If field research is itself a kind of camping practice, then these sites of research leave important historical traces of ephemeral occupations of place—perhaps more so than the nonextant camping interventions.

43. Marion Post Wolcott's documentation will serve as the basis for a subsequent narration of a walk through the Sarasota City Trailer Camp.

44. Ignasi de Solà-Morales, "Mediations in Architecture and in the Urban Landscape," in *Cities in Transition,* ed. Arie Graafland and Deborah Hauptmann (Rotterdam: 010 Publishers, 2001), 284. In this excerpt, de Solà-Morales echoes Jean Baudrillard's similar contentions about the construction of the photograph. Also, note James Agee's identification of photograph with fact, or at least more fact than prose can be. Also, note Harbison's portrayal of the "immediacy of the fragment"—Agee's text and his cataloguing procedures hold together the most when they are the most fragmented. Robert Harbison, *Eccentric Spaces* (New York: Avon, 1980), 154–62. This idea also relates to the discussion of the role of the catalogue/manual/map.

45. This set of rules comes from bylaw section 1.3, dating from the 1920 copy in the Florida State Archives in Tallahassee.

46. Barnet Hodes and Gale Roberson, *The Law of Mobile Homes,* 3rd ed. (Washington, D.C.: Bureau of National Affairs, 1974), 105, quoted in Wallis, *Wheel Estate,* 71–72.

47. According to laws of the time, four hundred square feet was considered minimally adequate dwelling space.

48. Wallis, *Wheel Estate,* 76.

49. Refer to the "operational manual" later in this chapter.

50. See chapter 8.

51. Wallis, *Wheel Estate,* 73–74. For a complete discussion of this sequence of events and its grandfathered characteristics, refer to chapter 2, "Image and Invention," in Wallis's *Wheel Estate.*

52. According to the *Oxford English Dictionary* (2nd ed.), the French words *attacher* and *attaquer* are closely related (1: 758–59).

53. An early prototype for Curtiss's personal use was the Adams Motor Bungalow, which according to Elon Jessup was classified as the "palace car" type because of its cost and elaborate construction.

54. Edward Casey, *The Fate of Place* (Berkeley and Los Angeles: University of California Press, 1997), 302. Casey interprets Deleuze and Guattari's opposition of nomad (*dispars*) and state or royal sciences (*compars*). Deleuze and Guattari's discussion occurs in *Nomadology: The War Machine,* trans. Brian Massumi (New York: Semiotext[e], 1986), 31–34.

55. This situation can be found in Gibsonton's domestic sites of invention discussed in chapter 8.

56. Henry David Thoreau, *Walden* (Princeton: Princeton University Press, 2004), 57. A comparison might be made between Thoreau's house-tent crystallization and Michel Butor's description of the Turkish canopies becoming rigid material over time.

57. Elon Jessup, *The Motor Camping Book* (New York: Putnam's, 1921), 141.

58. This type also includes the Des Moines double tent, the lean-to with sewed-in floor, and the faceted wedge-style, with open front to face the campfire. Jessup discusses these types throughout the text. Ibid., 123, 120, 122, 126–28, and 151.

59. Jessup's running-board boxes literally open up to allow for storage and for use as tables and stove. See the "Boxing" section.

60. The *Oxford English Dictionary* (2nd ed.) includes a primary denotation of "spreading out on the ground."

61. Jessup, *The Motor Camping Book,* 132.

62. Ibid., 78. Jessup continues, "The left-side tonneau door was permanently closed while traveling but both front seat doors were free."

63. Ibid., 79. "In the lower space under the second shelf were packed a gasoline stove and the cooking utensils."

64. Ibid., 82–83. This passage is reminiscent of a Homeric catalogue such as the listing that occurs at the beginning of the *Iliad.* Jessup continues in detail: "Various suit-case trunks used in motoring are of the same general lines and very likely one of these was the inspiration for the idea. The front is in two lengthwise sections, one hinged to the top of the box and the other to the bottom so that in opening one flanges down and the other up. The box is made of three-ply basswood and the addition of a covering all around of enameled duck keeps out dust and rain. The interior is partitioned into two main compartments, the larger of these being as high as the box and thirty-three inches long. In this, snugly fits a large sized suit case containing clothing and other personal effects. Packed above the suit case is an air mattress. The remaining seventeen inches of lengthwise space in the box is devoted to culinary matters. This is subdivided into five smaller compartments. In the upper left-hand corner is packed a coffee pot, bacon, and ham. The one next to it holds six one pound boxes of coffee, sugar, flour, meal, rolled oats, and rice. In the left middle compartment are canned goods of various sorts—milk, soups, Crisco, and so on. The right middle space is given over to a fifty-one piece set of nested aluminum ware such as I described in Chapter V. In the lower compartment next to the running board first a two-burner gasoline stove. . . . With this equipment, the owner and his family covered more than six thousand miles during one summer and fall."

65. Thoreau, *Walden,* 24. Thoreau also notes: "I have seen Penobscot Indians, in this town, living in tents of thin cotton cloth, while the snow was nearly a foot deep around them. . . . Formerly, when how to get my living honestly, with freedom left for my proper pursuits, was a question which vexed me even more than it does now, for unfortunately I am become quite callous, I used to see a large box . . . and, having bored a few auger holes in it, to admit the air at least, get into it when it rained and at night, and hook down the lid, and so have freedom in his love, and in his soul be free. This did not appear the worst, nor by any means

a despicable alternative. . . . Many a man is harassed to death to pay the rent of a larger and more luxurious box who would not have frozen to death in such a box as this."

66. The house-cars of the Tin Can Tourists also mixed the volumetrics of the box with the freedom of modification. One house-car, owned by M. Hoover of Ohio and documented in Arcadia, Florida, in 1929, was mounted on a Model T one-ton truck with the added amenities of a thirty-gallon water tank, a veranda, and an icebox. The "Harriet" house-car, owned by Harriet Warren, Flora Kavanaugh, and Westel Ashe of Brattleboro, Vermont, included a continuous clerestory light and operable bays that could open up for relaxing in wicker chairs. (Photographs of many house-cars, in particular these versions photographed in Arcadia, Florida, at the Tin Can Tourist Convention, January 10, 1929, are in the collection of the Florida State Archives in Tallahassee.)

67. Thoreau, *Walden*, 29, 113.

68. These mobilized homes include recreational vehicles, travel trailers, and autocampers.

69. Wolcott's full caption for the woman sitting next to her shell garden is "Guest at Sarasota trailer park, Sarasota, Florida, beside her garden made of shells and odds and ends. The camp has a garden club for members." See also the illustration on p. xviii and the discussion in chapter 8.

70. "Nomadic hesitation" is at odds with both the siting of camp, which is a conflict between place and space, and the timing of camp, which is a conflict between temporality and duration. See additional discussion of Deleuze and Guattari's term in chapters 5 and 11.

71. Refer to the discussion of Michel de Certeau's terminology "practiced place" in chapter 2.

72. Frank Wing, "An Elegy in De Soto Park," published in the *Tampa Morning Tribune,* April 4, 1924. The poem was published one month after the Tin Can Tourists were officially evicted from De Soto Park.

73. Krauss's primary focus is contemporary high art's intervention into the urban landscape as opposed to the "low" bricolaged art of tourism. See Rosalyn Deutsche, *Evictions: Art and Spatial Politics* (Cambridge: MIT Press, 1998), xv.

74. Friedrich Nietzsche, "Homeless" ("Ohne Heimat"), 1859, trans. R. J. Hollingdale, in *Nietzsche: The Man and His Philosophy* (1965). Nietzsche's definition of the homeless here is clearly divergent from that of Krauss, but his sentiments about mobility and perceptions of mobility provide a transition to the discussion of itinerancy and place.

75. E. B. White, "On a Florida Key," in *Essays of E. B. White* (New York: Harper and Row, 1977), 140. See chapter 10 for a discussion of the "unfinished cities" of camping.

CHAPTER 7. BRADEN CASTLE PARK

1. Ed Genia, D. L. Barlow, and R. W. Vaughn.

2. It is unclear from archival documents the extent to which direct conflict with Ybor City's Cuban and Cuban American population precipitated the expulsion.

3. The committee included W. J. Houck, president; L. K. Supernaw, vice president; R. W. Vaughn, secretary; and W. B. Jacobs, treasurer. Membership fees were fixed at one dollar for

the family, with fifty cents in annual dues. Shares cost ten dollars, with a limit of fifty shares per person. See R. W. Vaughn, "Origin and History of the Camping Tourists of America," Braden Castle Association, dated January 8, 1927, Eaton Florida History Room, Manatee County Central Library, Bradenton, Fla.

4. Manatee County, Clerk of Circuit Court, Incorporation Book C: 335–37, Bradenton, Fla.

5. See Loren O. Binkley and Pamela Gibson, "Braden Castle Camp and the Camping Tourists of America," April 1984, typescript, Eaton Florida History Room, Manatee County Central Library, Bradenton, Fla.

6. On March 9, the purchase of the land was officially made.

7. Bradenton Chamber of Commerce, "Bradenton Florida Tourist Club" (unpublished pamphlet, 1929–30), Eaton Florida History Room, Manatee County Central Library, Bradenton, Fla.

8. On April 3, 1924, Robbins was appointed historian and was directed to write a history of the castle property. Camping Tourists of America Minutes, 16.

9. H. E. Robbins, "History of Braden Castle, Florida: A Soliloquy by the Old Castle in History and Romance—A True Story," typescript in the collection of the Braden Castle Association, Bradenton, Manatee County, Fla. (n.p., n.d.).

10. Tabby construction combines local materials of lime, sand, and oyster shells.

11. Paul Eugen Camp, "The Attack on Braden Castle: Robert Gamble's Account," typescript, Eaton Florida History Room, Manatee County Central Library, Bradenton, Fla., 2. Other sources place the completion of the castle earlier; for example, Jack B. Leffingwell cites 1843 as the completion date to coincide with reports of a prolonged attack by Seminole Indians in the area who, according to some accounts, sacked the castle's grounds. "Fort Braden: Beginning of Bradenton Was Part of a Venture of Braden Brothers and Making of Sugar," typescript, Eaton Florida History Room, Manatee County Central Library, MM 18C (n.p., n.d.).

12. Robert Gamble, "Some Recollections of the Seminole Chief Arpioka—Bowlegs—and His War with the States," Richard Keith Call Papers, Florida Historical Society, University of South Florida (n.p., n.d.).

13. Postcards that were produced before the founding of Braden Castle Park (in the collection of the Manatee Country Clerk of the Court) portray Sunday picnics amidst the ruins.

14. Note that H. E. Robbins uses the phrase "in History and Romance" in his history of the castle.

15. Braden Castle Park was included as a historic district in the National Register of Historic Places in 1983 (83001428).

16. Elections were held on February 19 and 20; and in the next meeting on February 25, 1924, the officers decided to make the advertisement.

17. The Camping Tourists were hosted by the real estate firm Sharp and Root, which had been recommended by the Bradenton Chamber of Commerce.

18. Minutes, Braden Castle Association, 7.

19. See the section of the Minutes dated March 9, 1924, but appearing out of sequence in the text, 22–25.

20. Ibid., February 20, 1924, 5.

21. See entries in ibid., March 22 and 27, and April 1, 1924.

22. Ibid., April 1, 1924, 14. The next section of this chapter will explore this idea in terms of the utopian construct, particularly for its resonance with Thomas More's framing of "utopia."

23. Ibid., April 3, 1924, 16.

24. Ibid.

25. It appears that the officers had in mind third parties who would be renting the trailers for extended periods of time.

26. Ibid., April 5, 1924, 19.

27. Ibid., April 7, 1924, 20.

28. Ibid., 20–21.

29. R. W. Vaughn, "Origin and History of the Camping Tourists of America." In the text, Vaughn continues: "but [they] did make many promises to us, some of which, sorry to say, have not been kept."

30. See "Braden Castle, Tourist Center, Is a Part of Manatee," copied from Manatee Chamber of Commerce Celebration Edition and rewritten from a feature article prepared for Braden Castle Camp in January 1926. The manuscript is included in the collection of the Braden Castle Association.

31. Ibid.

32. Entry for "bylaw," in the *Oxford English Dictionary*, 2nd ed., 20 vols., prepared by J. A. Simpson and E.S.C. Weiner (New York: Oxford University Press, 1989). In an early form, the term *bymenighed* was used in Scandinavian languages to denote an association of farmers, commonly occupying a rural township.

33. Ibid., 2: 734.

34. More's *Utopia* was first published in 1516.

35. "Braden Castle, Tourist Center, Is a Part of Manatee."

36. The connection between Thomas More's *Utopia* and the bylaws can also be seen in the treatiselike quality of the earlier text. The lengthy title of More's work characterizes the project as a "handbook": "On the Best State of a Commonwealth and on the New Island of Utopia: *A Truly Golden Handbook*, No Less Beneficial than Entertaining. . . ." (italics added). Indeed, many subsequent planners and architects such as Sant'Elia, Tony Garnier, and Patrick Geddes have taken up this project.

37. In this model, religion is regarded as a necessary social institution (Thomas More, *Utopia*, trans. Robert M. Adams (Cambridge: Cambridge University Press, 1995), 161, 219, 223). In Braden Castle Park, religious and political requirements are integrated into the early drafts of the bylaws (see Minutes, April 1, 1924, 14). The most explicit reference to this assimilation is found in article XI § 2, prohibiting manual labor and games anywhere in the park on Sunday.

38. It should be noted that there is a marked difference in scale between the thirty-four-

acre park and the nearly two-hundred-mile-wide imagined island of Utopia in which multiple cities can be found.

39. More, *Utopia*, 109.

40. Ibid., 111, 119.

41. The camp at Braden Park evolved from a temporary settlement similar to its earlier incarnation within De Soto Park to the more permanent, regulated community of Braden Castle Park. The minutes generated from preliminary meetings of the newly formed Camping Tourists of America describe how the camp was initially conceived and organized; a chronological review of these proceedings reveals the transformation from temporary to permanent camp. From its inception, Braden Castle Park was designed to include three types of housing, with varying degrees of permanence and of size. These categories of dwelling are the campsites identified as the "trailer camp," ten sites for cottages, and houses within the main grounds of the park. Refer to H. E. Robbins's drawing of the grounds (1936) for the layout of these types in his "History of Braden Castle, Florida."

42. Constitution and By-Laws of Camping Tourists of America, Braden Castle Park, Bradenton, Manatee County, Fla., 1945, art. IX § 1.

43. Ibid., art. IX § 2.

44. Ibid., art. IX § 3.

45. More, *Utopia*, 121.

46. Constitution and By-Laws, art. IX § 4.

47. The original number of properties was set at two hundred based on the maximum membership allowed for by the bylaws (art. II § 3). The National Register of Historic Places also notes that three additional structures (the clubhouses and community hall) make up the historic site.

48. Vaughn, "Origin and History of the Camping Tourists of America," 2.

49. "In the Land of Manatee," Braden Castle Association. Artisans as well as doctors and lawyers resided at Braden Park.

50. More, *Utopia*, 133. More notes the following about tradespeople in Utopia: "each person is taught a particular trade of his own, such as wool-working, linen-making, masonry, metal-work, or carpentry. . . . Every person learns one of the trades" (125).

51. Constitution and By-Laws, art. III § 4.

52. Ibid., art. IV § 1. Section 2 notes that the directors "shall select from their own members a President, a Vice President, a Secretary and Treasurer." Here, the mode of government of the park diverges from that of Utopia. Each group of thirty households elects a syphogrant to represent them (More, *Utopia*, 121). Although also elected by a "secret vote," the voting for the highest officials is not carried out by the general populace and is instead conducted by the syphogrants (ibid., 131). In general, public officials of More's Utopia must be scholars; and it is interesting to note that three of the nine original officers of the Camping Tourists organization were identified by the title "Dr." Research has been unable to determine if these titles refer to medical or academic degrees.

53. Constitution and By-Laws, art. X § 1; italics added.

54. Ibid., art. XI § 1, 2, and 3. For the origin of this resolution, see the discussion in the previous section of this chapter and in the association's minutes, April 1, 1924, 14.

55. Robbins, "History of Braden Castle," 16. A parallel between More's work and Robbins's narrative history is that the 1516 publication of *Utopia* includes two introductory inscriptions that personify the island, which narrates its story. One of these is excerpted later in this section, and the other served as an example of the Utopian language. Titled "A Quatrain in the Utopian Language," this latter text reads: "The commander Utopus made me, who was once not an island, into an island. I alone of all nations, without philosophy, have portrayed for mortals the philosophical city. Freely I impart my benefits; not unwillingly I accept whichever is better" (23).

56. More, *Utopia,* 247. More continues in a passage resonating with the earlier conflict of the Tourists with the neighboring municipalities and with their search for home: "As long as they preserve harmony at home, and keep their institutions healthy, they can never be overcome or even shaken by all the envious princes of neighboring countries, who have often attempted their ruin, but always in vain" (247).

57. See note 10.

58. On the facing page of this introductory passage appears an advertisement for Wakeman Funeral Home in Bradenton.

59. Robbins, "History of Braden Castle, Florida," 1.

60. In *The Tourist* (1989) and the subsequent *Empty Meeting Grounds: The Tourist Papers* (New York: Routledge, 1992), Dean MacCannell examines the possibility of tourism serving as the "primary ground for the production of new cultural forms on a global base." In the introduction to the latter publication, MacCannell notes, "In short, tourism is not just an aggregate of merely commercial activities; it is also an ideological framing of history, nature, and tradition; a framing that has the power to reshape culture and nature to its own needs" (1). For this tourist who, as the neonomad, constructs "new hybrid cultural forms" and must rely on stories accumulated from his or her travels, "the only reality is the imagination" (2, 4).

61. Robbins, "History of Braden Castle, Florida," 16.

62. This outlook is also found in John Nolen's *New Towns for Old,* which was published in 1927—thus overlapping with the Braden Castle Park project. At the conclusion of his work, Nolen also addresses the idea of creating a new community: "The new order of community life such as is here roughly and briefly depicted as being possible in the planning of satellite towns . . . ought to include more things that make life worth living: decent homes, . . . fit bodies and active minds; . . . reasonable quiet; and, above all, safety from danger and disease. In these new cities we could . . . add much to the decoration and adornment of life and its legitimate amusements and recreations. . . . Indeed by building anew we could raise the whole plane and standard of the common life." *New Town for Old* (Boston: Marshall Jones, 1927), 156–57.

63. Robbins, "History of Braden Castle, Florida," 16.

64. It could be said that the Tourists of Braden Castle reunite the Wanderer and the Cave-Dweller of Frank Lloyd Wright's "Disappearing City." The radicalized mobility of the Wanderer is made partially static by the sedentary existence of the Cave-Dweller. Wright himself sought to find a model that would allow for the two antithetical tendencies to coexist. The "radical decentralization" of his Broadacre City would, according to Wright, create an "organic social order" in which freedom (of individualism) could mesh with the conservatism

of greater society. See Robert Fishman, *Urban Utopias in the Twentieth Century: Ebenezer Howard, Frank Lloyd Wright, Le Corbusier* (Cambridge: MIT Press, 1982), 156–58. Wright's ideal city seeks this coexistence through decentralization and dispersal while Braden Castle Park is framed as a "city within a city" and as a densely woven compact city of modest cottages. It is, however, not surprising that the plan for Broadacre City includes a Tourist Camp.

65. "Six Lines on the Island of Utopia written by Anemolius, Poet Laureate, and Nephew to Hythloday by his Sister." More, *Utopia*, 19.

66. Robbins, "Braden Castle," in the front matter of "History of Braden Castle."

CHAPTER 8. GIBSONTON

1. Originating from the particular situations found in Gibsonton, Florida, the idea of camp as parasite is used in this chapter to address the multivalent and often paradoxical relationship of place, time, and region. The term is introduced not for its distinctly negative connotations of a biological organism, but for its operational tendency (as *para-site*) of simultaneously maintaining and transforming within a host-guest relationship. This basic idea has its origins in Michel Serres's treatment of the term in his work entitled *The Parasite,* in which Serres begins his theory of human relations with the premise that "parasite" is host, guest, and noise. Even more important to this discussion is Serres's contention that the parasite attaches to the *relation* between host and guest, basically "parasiting" the host indirectly through an existing system. Recent work on parasitism has also been carried out by Jacques Derrida in his study of hospitality and by Andrew Benjamin in "Parasitism in Architecture"—an introductory theoretical framework provided for the international competition to design ephemeral structures for the 2004 Olympics in Athens, Greece. Under this general theme, three iterations of the parasitic mode will be investigated through the Gibsonton campsite: parasite as excluded middle, parasite as system of relations (in which the parasite is both a relational term and an instrument of change), and parasite as duration. See Michel Serres, *The Parasite,* trans. Lawrence R. Schehr (Baltimore: Johns Hopkins University Press, 1982); Jacques Derrida, *De l'hospitalite autor de Jacques Derrida* (Passe du Vent: Genouilleux, 2001); and Andrew Benjamin, "Parasitism in Architecture," in *Ephemeral Structures in the City of Athens: International Competition Program*, 55–62 (Athens: Hellenic Cultural Heritage SA, 2002).

2. While its name and nicknames both include "town," this particular place does not have the distinct boundaries or governing apparatus that denote a town. And its population exceeds what might typically define a village but presumably does not constitute a city (if not for its lack of political definition, then for its decidedly suburban texture). As an unincorporated "census-designated" place, Gibsonton does have a measurable population. Based on the 2000 Census boundaries and survey, 8,752 live in the zone, with a population density of 681 people per square mile. In spite of this population growth, Gibsonton maintains its original character, described in the Federal Writers' Project's *Florida: A Guide to the Southernmost State* (New York: Oxford University Press, 1939) as "a small trailer camp and filling station on the southern bank of the Alafia River [that] was named for the pioneer Gibson family" (403).

3. Judy Tomaini, interview by the author, January 28, 2003, Gibsonton, Fla.

4. Ibid.

5. The lease was set at one dollar per year in the 1940s for approximately thirty feet of bayfront property.

6. Tomaini interview, January 28, 2003.

7. The setting of Gibsonton allows the performer-residents to choose the degree of public life during the winter season respite. As a de facto community center, Giant's Camp Restaurant allows for publicity within Gibsonton's self-regulated public sphere.

8. Airing March 31, 1995, "Humbug" was episode twenty in season two of the *X-Files*.

9. The Alligator Man is a character loosely based on Grady Stiles, known as Lobster Boy. Many of the carnival performers in Gibsonton, including the Stiles family, resonate with the archetypal characters in Katherine Dunn's *Geek Love* (1993).

10. This nominal change is interesting though not necessarily significant in terms of the argument for an ambiguous and labile character. "Conundrum" denotes punning or wordplay, while "enigma" suggests a more deeply complex riddle and as such a more ingrained possibility of multiple readings or misreadings.

11. By 1926, Highway 41 extended to a southern terminus in Naples; and in 1949, Highway 94 was added to the route linking Naples and Miami as a part of the "Tamiami Trail" connecting Tampa with the east coast across the Everglades. While not as identifiable as Routes 1 and 66, the highway has entered the American imagination with such references as the Allman Brothers' song "Ramblin' Man" from the 1970s.

12. The distance from Miami to Valdosta is approximately 450 miles.

13. Fred Kniffen, "The American Agricultural Fair: The Pattern," *Annals of the Association of American Geographers* 39, no. 4 (December 1949).

14. A story told in the Federal Writers' Project's *Florida: A Guide to the Southernmost State* also suggests that this area has been the site of buried treasure in regional folklore. See the discussion of Gibsonton in chapter 5.

15. It is interesting to note that the *campo* is reintegrated into the urban fabric as the open square (the country in the city).

16. In ancient Greek, uses of the term *chora* and its variants can be lexically connected to campsites. In the *Iliad*, for example, Hector makes camp in a "space cleared of the dead"—an open space (*choros* as χωρος) that is outside the walls and the jurisdiction of the *polis*. Homer, *Iliad*, trans. A. T. Murray (Cambridge: Loeb Classical Library, 1999), 8.489.

17. Allan D. Wallis, *Wheel Estate: The Rise and Decline of Mobile Homes* (New York: Oxford University Press, 1991), 39.

18. The transformation of the excluded middle (*mi-lieu*) into the *milieu* has implications for how we understand place and, for that matter, region. Embedded within the French language, this distinction involves a potential movement from the local to the global: "New reversal: from the half-place [*mi-lieu*], a small excluded locality, insignificant, ready to vanish, to the milieu [*milieu*], like a universe around us." Michel Serres, *The Troubadour of Knowledge*, trans. Sheila Faria Glaser and William Paulson (Ann Arbor: University of Michigan Press, 1997), 43.

19. See, in particular, the following works by Michel Serres: "The Third Man" in *The Troubadour of Knowledge*, 46–48; "Decisions, Incisions: The Excluded Third, Included" in *The Parasite*, 22–25; and "The Birth of Time" in *Genesis*, trans. Geneviève James and James Nielson (Ann Arbor: University of Michigan Press, 1995), 81–122.

20. Serres, *The Parasite*, 16. For previous work relating the parasite and the midway, see Robert Segrest's "The Perimeter Projects: Notes for Design," *Assemblage* (1986): 24–36; and his subsequent "The Perimeter Projects: The Architecture of the Excluded Middle," *Perspecta 23: Yale Architectural Journal* (1987).

21. Carla I. Corbin, "The Old / New Theme Park: The American Agricultural Fair," in *Theme Park Landscapes: Antecedents and Variations*, ed. Terence Young and Robert Riley (Washington, D.C.: Dumbarton Oaks Research Library and Collection, 2002), 197.

22. Don Gregory (midway manager for the Florida State Fair), interview by the author, Tampa, Fla., February 1, 2003.

23. This recalibration of public and private in camps parallels confluences of houses and new media. The introduction of the television set into the modern house suggested a new reading of what in domestic life was hidden and revealed. The understanding of the house as theater created a contested ground between private viewing and a viewing public. As Beatriz Colomina notes, the exhibitionism of the modern house reveals an interior that is already in many ways prepared for its exposure. As in Gibsonton in particular and many camps in general, the living room of the modern house mixes public and private. In more contemporary electronic media, this idea of exhibition reaches its culmination: "The modern house has been deeply affected by the fact that it is both constructed in the media and infiltrated by the media." "The Media House," in *The City of Small Things* (Rotterdam: Stichting Parasite Foundation, 2000), 117. The blurring of public-private characteristics of camps holds possibilities for understanding the effect of such a wholly *mediated* experience of living spaces, because camps already rely on a high degree of exposure for the *mediation* of the private through the public. For the discussion of the relation between modern house and media, see Colomina, "The Media House."

24. While not necessarily functioning as an actual place of performance, the tacit and ad hoc social contract of symbolizing the reconfigured apparatus and the spaces of the midway is a phenomenal presence in Gibsonton.

25. John Brinckerhoff Jackson, *Discovering the Vernacular Landscape* (New Haven: Yale University Press, 1984), 86.

26. This idea of the vernacular as a "living organism" is derived from Dante's *De vulgari eloquentia*. This concept, along with vernacular as process, is discussed in greater detail in chapters 4 and 12.

27. Jackson, *Discovering the Vernacular Landscape*, 151.

28. Hillsborough County Comprehensive Plan, Sec. 3.01.02 SB—Show Business Overlay District/Purpose.

29. Ibid.

30. Hillsborough County Comprehensive Plan, Sec. 3.01.

31. This definition is taken from the U.S. Census Glossary found at http://factfinder.census.gov/bf/_lang=en_vt_name=DEC_2000_SF1_U_DP1_geo_id=16000US1225900.html.

32. Segrest, "The Perimeter Projects: The Architecture of the Excluded Middle," *Perspecta 23: Yale Architectural Journal* (1987): 54–55.

33. In *Reading with Michel Serres: An Encounter with Time* (Albany: State University of New York Press, 1999), Maria Assad notes about the Serresian parasite, "the excluded third insinuates . . . [itself] into a given system only to become, in turn, the system per se" (21).

34. Claude Levi-Strauss, *The Savage Mind* (Chicago: University of Chicago Press, 1966), 16. Levi-Strauss continues: "In its old sense the verb '*bricoler*' applied to ball games and billiards, to hunting, shooting and riding. . . . And in our own time, the '*bricoleur*' is still someone who works with his hands and uses devious means compared to those of a craftsman." Levi-Strauss's characterization of the *bricoleur* is similar to Michel Serres's treatment of the figure of Hermes as both relay and operator for bringing "things" together. Hermes, like the *bricoleur,* works *between.* An exchange between Bruno Latour and Michel Serres helps to clarify the figure of Serres's Hermes, particularly in terms of the *bricoleur*'s seemingly "devious" methodology. Moving and traveling in this excerpt can be considered analogous to the question of making in Levi-Strauss's discussion. Latour begins, "If we return to the problem of traveling from place to place, which is thus the problem of 'folding' time—effecting juxtapositions—and the problem of metaphor." Serres replies, "Metaphor, in fact, means 'transport.' That's Hermes' very method: he exports and imports; thus, he traverses. He invents and can be mistaken—because of analogies, which are dangerous and even forbidden—but we know no other route to invention. The messenger's impression of foreignness comes from this contradiction." Michel Serres and Bruno Latour, *Conversations on Science, Culture, and Time,* trans. Roxanne Lapidus (Ann Arbor: University of Michigan Press, 1995), 66.

35. Serres, *The Parasite,* 160–61. Serres also notes: "The joker is a logical object that is both indispensable and fascinating. . . . Just as the most general model of method is the game, the good model for what is deceptively called bricolage is the joker."

36. See the discussion of J. B. Jackson's definition of the vernacular in chapter 4.

37. See Charles Jencks and Nathan Silver, *Ad Hocism: The Case for Improvisation* (New York: Doubleday, 1972).

38. Refer to Bergson's concept of the "present that was" and the "past that is" in Deborah Hauptmann, "The Past Which Is: The Present That Was," in *Cities in Transition,* ed. Arie Graafland and Hauptmann, 350–61 (Rotterdam: 010 Publishers, 2001).

39. See, for example, the method circumscribed by David Greene in "Gardener's Notebook," in *Archigram* (New York: Princeton Architectural Press, 1999), 110–15.

40. According to the *Oxford English Dictionary* (2nd ed., 20 vols., prepared by J. A. Simpson and E.S.C. Weiner [New York: Oxford University Press, 1989]), "yard" derives from the Old Saxon *gard* (enclosure, field), linked to the Dutch *gaard,* meaning "garden" and distantly related to Latin *hortus.* In contemporary usage, yard also refers to a kitchen or cottage garden—a garden closely linked with and scaled to domestic production.

41. This "signage" occurs both semiotically in the indexical function of the shell gardens photographed by Marion Post Wolcott and literally in the signs that identify performers' homes and that celebrate their itinerancy.

42. As Wolcott's extensive caption notes, Sarasota City Trailer Camp offered a garden club for those campers wintering there. See illustration on p. xviii.

43. The trailer-hitch garden can be compared to the blocking details of contemporary mobile homes that have become chronically immobilized.

44. On Gilles Deleuze's concept of bricolage, Elizabeth Grosz writes: "the subject . . . is a series of fragments or segments capable of being linked together in ways other than those that congeal it into an identity. 'Production' consists in those processes that create linkages between fragments—fragments of bodies and fragments of objects—and 'machines' are heterogeneous, disparate, discontinuous assemblages of fragments brought together in conjunctions (x plus y plus z) . . . a concept not unlike a complex form of bricolage or tinkering described by Levi-Strauss. . . . They are multiplicities of (more or less) temporary alignments of segments." "A Thousand Tiny Sexes: Feminism and Rhizomatics," in *Gilles Deleuze and the Theatre of Philosophy,* ed. Constantin Boundas and Dorothea Olkowski (London: Routledge, 1994), 198. As a bricolaged assembly, the shell garden becomes an organic machine of production and symbolization.

45. Michel Butor, "Bricolage: An Interview," by Martine Reid and Noah Guynn, *Yale French Studies* 17, no. 10 (January 1994): n. 84.

46. Camps, particularly those that incorporate assemblages such as the shell gardens, become memory theaters of travel. Such camps relate to the combination of narrative and temporality present in Chinese travel gardens. In the context of Florida's surreal landscape, Nina Hofer has written of the Suzhou travel gardens in relation to the now-defunct theme park Splendid China near Orlando. Hofer notes that the Orlando version of the garden might provide a tourist's analog for "world-making" and "world recovery." "Letter from China," in "Urbanism versus Architecture," special issue, *ANY,* no. 9 (1994): 7–9.

47. In "Lot, Yard, and Garden: American Distinctions" (*Landscape* 30, no. 3 [1990]: 29–35), Paul Groth has linked yard spaces, specifically those defined as "lots," with the drawing of lots for land. Reducing the distinctions between terms (and spaces), Groth equates lot with division, yard with enclosure, and garden with contents. Historically in the mid- to late nineteenth century, with the growth of backyard culture, contiguous front yards participated in what Groth calls "open-space semantics" in which "whole streets became a new kind of space . . . open, flowing, parklike spaces" now associated with suburban estates (33).

48. George W. Ferris designed the original "observation wheel" for the World's Columbian Exposition in 1893. Centered along both major and minor axes, the wheel anchored the midway's spaces not only in its monumentality but also as an attraction with panoramic views of Chicago and its surroundings. Accordingly, the wheel was both functional and symbolic. The wheel represented the possibilities of combining recreation and leisure with industrial and manufacturing progress. In spite of its extraordinary dimensions and scale, the Ferris wheel proved to be an ephemeral archetype and was dismantled in early spring 1894. The mobilized wheel was reassembled twice before being sold for scrap metal: in the Hyde Park neighborhood of Chicago between 1895 and 1903 and at the Louisiana Purchase Exposition in St. Louis 1894.

49. "A symbol is a law or regularity of the indefinite future." Charles Sanders Peirce, *The Collected Papers,* ed. Charles Hartshorne and Paul Weiss (Cambridge: Belknap Press of Harvard University Press, 1960), 4: 448.

50. The midway-yard coupling becomes a yard-midway coupling.

51. J. B. Jackson has pointed out that "mobility and change are the key to the vernacular landscape, but of an involuntary, reluctant sort . . . an unending adjustment to circumstances." Jackson's vernacular homesteads, like the campsites of Gibsonton, result from "hesitation," or a temporary immobilization that typically endures. This duration, however, occurs within the fluctuations of time and thus allows a flexibility for adaptation and ultimately invention. *Discovering the Vernacular Landscape,* 151.

52. See "Vernacular" in ibid., 85–87.

53. David Morley, *Home Territories: Media, Mobility and Identity* (New York: Routledge, 2000), 4. Morley enlists Philip Tabor's "spaces of belonging" to talk about the territory of home.

54. For the most part in Morley's discussion, mobility is metaphoric. See the previous note.

55. Such a particularity of place falls within what Edward Casey has called "place-as-region." *The Fate of Place* (Berkeley and Los Angeles: University of California Press, 1997), 246, 305. For the performers living in Gibsonton, the seasonal network of carnivals and fairs serves as the region in which these particular reconstructions of the midway occur. The construction of the midway formulates the local operations that characterize the relation to place. Though arising from the model of the midway, the flexibility and fluidity of local operations allow for each place to be understood in its specificity and in its connotation as region.

56. Charles Peirce uses this phrase to describe the process of logic and symbolization.

57. Note the similarity to the "traveling-in-place" aspect of the collection of dirt housed at Giant's Camp.

58. Judy Tomaini, interview by the author, Gibsonton, Florida, February 1, 2003.

59. Ignasi de Solà-Morales, *Differences: Topographies of Contemporary Architecture* (Cambridge: MIT Press, 1997), 104.

60. Henri Bergson, *The Creative Mind: An Introduction to Metaphysics* (New York: Citadel, 1992), 188–91.

61. Henri Bergson summarizes this relation to experience: "Our concept of duration was really the translation of a direct and immediate experience." *Duration and Simultaneity: Bergson and the Einsteinian Universe,* trans. Leon Jacobson (Manchester: Clinamen, 1999), xxvii.

62. This idea is closely related to what Peirce refers to as the "nature of symbols" and characterized by the idea proposed here of symbol as process.

63. Celebration resident Paul Cooney notes the feelings elicited upon seeing the fence at the edge of his yard in the New Urbanist community: "I always feel an enormous sense of relief whenever I see that white fence. . . . When I see that fence, I know I'm home. It's like my heart rate slows and even my breathing becomes more relaxed." Quoted in Michael Lassell, *Celebration: The Story of Town* (New York: Roundtable Press, 2004), 36. Cooney's statement about the "sense of relief" at the sight of his fence provides an interesting example of what Bergson understands to be the confusion of "feelings," essentially solidified impressions to produce language, with the "permanent external object," or even "the word that relates to,

identifies, or expresses the object." Bergson, *Duration and Simultaneity*, 237–40. In Gibsonton, these associations (in the process of symbolization) are made through and within the midway-yard spaces.

64. Henri Bergson, "Introduction to Metaphysics," in *The Creative Mind* (New York: Citadel, 1992), 188.

65. The mobility of home becomes a tangible space that can be experienced through a particular place. If change of state is the rule, then the itinerancy of possibilities (possible worlds, yards, and homes) makes for a very subtle experience of place in a complex of minor events related to the archetypal camping procedure (siting, clearing, making, and breaking). These momentary place-ments then become immediate symbolizations of home. We experience the event of home as the gray area (Bergson's "indefinite") between fixed substantives and fleeting appearances or symbolizations.

CHAPTER 9. SLAB CITY

1. Dorothy Ann Phelps points out that thirty acres near the campsite had been cultivated in the 1930s but was never planted. "A Singular Land Use in the California Desert," master's thesis, University of California at Riverside, 1989, 40.

2. Technical Report and Project History, Contract No. 5426, prepared by Kister, Curtis, and Wright Architects and Engineers, San Diego, Calif., March 30, 1944.

3. The camp was named after Brigadier General Robert H. Dunlap, who served in the U.S. Marine Corps during World War I.

4. The tanks, to the west of Slab City, still exist and can be seen in aerial photography.

5. Because the manhole covers have been removed over time, the State Lands Commission has asked the Niland fire department to cover the open holes. However, this project does not appear to have been completed. Phelps, "A Singular Land Use," 44.

6. The North African war theater includes instances of the Roman *castrum*, particularly Timgad, which lies in an active World War II military area—the Algerian–French Morocco zone between 1942 and 1943. Phelps notes that artillery and anti-aircraft units of the Fleet Marine Force were trained at Camp Dunlap. Researchers in the past have sited General George S. Patton's headquarters at Camp Dunlap, but the actual site of Patton's camp is, according to the Department of the Interior (1986), in the Mojave Desert near Chiriaco Summit. "A Singular Land Use," 51. Military security policies during World War II disallowed photographs inside the camp, so analysis and evidence must come from textual documents, maps, and surveys.

7. Phelps notes that the temperatures for July 1942 averaged in the mid-110s, with a high temperature of 118 that month (ibid., 49). Water causes this type of soil to expand dramatically, and Phelps notes that in one instance a buckled slab was raised eight inches. Ibid., 50.

8. The diagram (titled *Diagrammatic Layout of a Tent Camp*) discussed here was included in FM 101–10, which was published by the U.S. War Department on October 10, 1943. This *Staff Officers' Field Manual* was produced primarily for the U.S. Army; but for the purposes of comparison and because of the diagram's apparent universality and generality, it is assumed that this model was to be found in other divisions of the military. The full title of the map is

"Map of Camp Dunlap, Niland, California, Marine Corps Artillery Range, Eleventh Naval District, San Diego, California, Showing conditions on June 30, 1944," and the map is drawn to a scale of 1 inch = 1,000 feet.

9. The Treaty of Guadalupe Hidalgo (1848) located the U.S.–Mexican border near its present longitude cutting across the deserts of southern New Mexico, Arizona, and California. Surveying errors resulted in a disputed land area known as the Mesilla Valley Strip and required the United States to purchase this zone in the Gadsden Treaty (1853). After Barlow and Blanco's survey and after 258 granite monuments replaced the shifting rock piles, the formation of a "no-man's-land" between the United States and Mexico at the active border crossing of Nogales completed the evolution of the American frontier space. The treaty also disallowed the revision of the international border as surveyed and designated with monuments by Emerson and Salazar in the 1850s. With the subsequently rapid improvement of surveying equipment, imprecision in the location of the treaty's mandate along the parallel 31° 47' was shown by subsequent surveys, most notably by Barlow and Blanco (1894).

10. Finally, President Theodore Roosevelt extended the zone, or "no-man's-land," as a sixty-foot-wide strip the entire length of the border from El Paso to the Pacific Ocean (35 *Stat.* 2136). This easement became known as the "Roosevelt Reservation." See Blas Nuñez-Neto and Stephen Viña, *Border Security: Barriers along the U.S. International Border,* Congressional Research Service report for Congress, December 12, 2006.

11. Eric Amptmeyer, response to Daniel Adkins, December 26, 2002, www.slabcity.org.

12. W. H. Auden, *Complete Works of W. H. Auden,* ed. Edward Mendelson (Princeton: Princeton University Press, 1993), 688.

13. De Certeau notes that this mediating role does not have the "character of a nowhere that cartographical representation ultimately presupposes." Instead, for de Certeau, stories rather than space begin to delimit frontiers. *The Practice of Everyday Life,* trans. Steven Rendall (Berkeley and Los Angeles: University of California Press, 1988), 127.

14. John Reps, *Cities of the American West: A History of Frontier Urban Planning* (Princeton: Princeton University Press, 1980).

15. The title of Turner's lecture at the 1893 Exposition was "The Significance of the Frontier in American History." See also Reps, *Cities of the American West,* x.

16. Reps, *Cities of the American West,* x.

17. Ibid., xii.

18. Robert Walter Frazer, *Forts and Supplies: The Role of the Army in the Economy of the Southwest, 1846–1861* (Albuquerque: University of New Mexico Press, 1983), xi.

19. Horst De la Croix, *Military Considerations in City Planning: Fortifications* (New York: Braziller, 1972), 35.

20. Ibid., 36.

21. The inner wall was made up of embedded elements, smaller towers, and fortified buildings.

22. Treatises on the art of war have echoed this idea, including William Garrard's *Arte of Warre* (1591), Sun Tzu's *Art of War* (written ca. 500–300 BC.), Machiavelli's *Art of War* (1520), and Karl von Clausewitz's *On War* (1832).

23. De Certeau, *The Practice of Everyday Life,* 35–36.

24. This type of construction recalls Deleuze's use of "nonlimited locality" to describe the nomadic place.

25. De Certeau, *The Practice of Everyday Life,* 36–37.

26. As a Cartesian approach, strategy transforms places into readable spaces—an activity that can be compared to the imposition of the "power of knowledge" by modern science and military strategy onto situations. Foucault draws a similar conclusion about power and knowledge from his reading of military camps. For Foucault, the architecture of the military camp reproduces and reclassifies knowledge and power through its hierarchy and regimentation: "the military hierarchy is to be read in the ground itself, by the tents and the buildings reserved for each rank. It reproduces precisely through architecture a pyramid of power." *Power/Knowledge,* ed. Colin Gordon (New York: Pantheon, 1980), 255. The tactic, on the other hand, is defined by de Certeau as "a calculated action determined by the absence of a proper locus."

27. De Certeau, *The Practice of Everyday Life,* 37.

28. Section 36, in which Slab City falls, was designated as a part of this grid system.

29. Nina Taunton takes a different view of these methods of fortification. Her analysis of the early treatises does not differentiate, in terms of weak and strong, the use of ditches-bastions from the formation of carriages to protect the camp. The authors that she draws from, particularly Fourquevaux's *Instructions for the Warres* (1589), justify the use of carriages as fortification in situations where the military campaign will not "tarry" for more than a day. See Taunton's citation of Fourquevaux's work in her "A Camp 'well planted': Encamped Bodies in 1590s Military Discourses and Chapman's *Caesar and Pompey,*" in *The Body in Late Medieval and Early Modern Culture,* ed. Darryl Grantley and Taunton, 83–96 (Burlington, Vt.: Ashgate, 2000), 85. This mobility also characterizes the occupation of Slab City by the "slabbers" in their trailers and RVs.

30. The full title of the land ordinance is "An Ordinance for Ascertaining the Mode of Disposing of Lands in the Western Territory." Passed by Congress on May 20, 1785, the complete entry for this allotment of land for educational purposes reads: "There shall be reserved the lot No. 16, of every township, for the maintenance of public schools, within the said township; also one-third part of all gold, silver, lead and copper mines, to be sold, or otherwise disposed of as Congress shall hereafter direct." http://www.cadcon.com/land_ordinance_of_1785.htm. Benjamin Hibbard notes a similar policy generated with the Land Grant Provision of the Ordinance of 1787, applying to the Northwest Territory the stipulation that section 16 be dedicated "to the use of schools." *A History of the Public Land Policies* (Madison: University of Wisconsin Press, 1965), 310.

31. The quadrant's full designation is read as "Section 36, T10N, R14E SBBM (San Bernardino Baseline and Meridian)."

32. Supplementary rules were proposed by the Environmental Protection Agency for the use of LTVAs in Arizona and California. Adopted in 2000 and 2002, these supplementary rules serve primarily as a clarification of the previously issued guidelines.

33. See "Notice of Proposed Supplementary Rules on Public Lands within all Arizona and California Long-Term Visitor Areas" (December 2002). See also "Publication of Supple-

mentary Rules for Long-Term Visitor Areas Managed by the California Desert District office, California, and the Yuma Field Office, Arizona." For rules issued in 2000 and specific to California areas, refer to *Federal Register* 65, no. 137 (July 17, 2000).

34. Phelps, "A Singular Land Use," 79.

35. Alex Necochea, e-mail to the author, April 8, 2003.

36. Larry Harvey, "The Burning Man—An Oral History," in *Burning Man*, ed. Brad Wieners (San Francisco: Hardwired, 1997).

37. In 2004, Burning Man organizers estimated attendance at 35,600 people.

38. According to Misrach's documentation, the Mobile Home attained a speed of 96 mph.

39. Refer to http://www.nv.blm.gov/winnemucca/recreation/Blackrock_advisory.htm.

40. The following is a typical coordinate location for a theme camp: 2,500 feet (Bowsprit) & 75 degrees. "Bowsprit" refers to the concentric ring that occurs between 2,500 and 2,700 feet from the Burning Man sculpture.

41. Guides offered at the Burning Man Web site include a resource guide and a survival guide: http://www.burningman.com/themecamps/resource_guide.html; http://www.burningman.com/themecamps/plan_village.html; and http://www.burningman.com/preparation/event_survival/2002_survival_guide.pdf.

42. Sarah M. Pike, "Desert Goddesses and Apocalyptic Art," in *God in Details: American Religion in Popular Culture*, ed. Eric Michael Mazur and Kate McCarthy (New York: Routledge, 2001). See also David Chidester, introduction to *American Sacred Space* (1995), 14, quoted in Pike, "Desert Goddesses," 157–58.

43. B. W. Gorham, *Camp Meeting Manual: A Practical Book for the Camp Ground* (Boston: H. V. Degen, 1854), 17, 65. Larry Harvey, who originated Burning Man, speaks of its sacred symbolism in an address he made in January 1997: "Liberty, at Burning Man, is tempered by our primal needs as human beings, and this shared experience, symbolized by our species' attraction to fire, forms a central and necessary basis for our community. A second basic lesson we have learned while acting in the abstract and liberating space of the desert is that in order to found a cultural sphere human beings require a center of gravity, a powerful axis in time and space. At our event this transcendent center is most conspicuously supplied by Burning Man himself."

44. Pike, "Desert Goddesses," 159. Pike derives this idea from Rob Shields, who writes, "an imaginary geography vis-à-vis the place-myths of other towns and regions which form the contrast which established its reputation as a liminal destination." *Places on the Margin: Alternative Geographies of Modernity* (New York: Routledge, 1991), 112.

45. Refer to http://www.nv.blm.gov/winnemucca/recreation/Leave_No_Trace.htm.

46. A part of what the Web site calls "Black Rock City Year Round," the main discussion board for Burning Man can be found at the E-Playa, http://eplaya.burningman.com/. Links from the main Web site also lead to a "2003 Lost and Found" page, http://www.burningman.com/blackrockcity_yearround/03_lostandfound.html.

47. These discussion groups include the bulletin board associated with slabcity.org and two Yahoo Groups, The Slabs and Rawhideboondockers.

48. Larry Harvey, "Burning Man and Cyberspace," from an address delivered January

1997. Harvey also notes, in terms of the cyberspatial quality of the festival, "Here it is possible to reinvent oneself and one's world aided only by a few modest props and an active imagination." http://www.burningman.com/whatisburningman/people/cyber.html.

49. The Internet serves as de Certeau's "mediating ground" between the distant participant and the place itself.

50. Henri Lefebvre has discussed the relation of time to space through the festival: "Thus time was not separated from space; rather it oriented space—although a reversal of roles had begun to occur with the rise of medieval towns, as space tended to govern those rhythms that now escaped the control of nature. . . . Where was the connection or bond between space and time? . . . In a praxis, an 'unconscious' praxis, which regulated the concordance of time and space by limiting clashes between representations and countering distortions of reality. Time was punctuated by festivals—which were celebrated in space." See *The Production of Space,* trans. Donald Nicholson-Smith (Oxford: Blackwell, 1991), 267. The festivals of Burning Man and the snowbird rituals of Slab City elide differences of time and place in a new space.

51. Hakim Bey provides on definition of the TAZ: "The TAZ [temporary autonomous zone] is like an uprising which does not engage directly with the State, a guerilla operation which liberates an area (of land, of time, of imagination) and then dissolves itself to re-form elsewhere/elsewhen, before the State can crush it." Slab City and Burning Man can be understood as subcategories of Bey's Temporary Autonomous Zone.

52. For Michel Foucault, heterotopias exhibit a proliferating set of relations. He could be describing the campsites on the Internet when he writes, in "Of Other Spaces" (*Lotus International,* nos. 48–49 [1986]): "I am interested in certain ones that have the curious property of being in relation with all the other sites, but in such a way as to suspect, neutralize, or invent the set of relations that they happen to designate, mirror, or reflect. These spaces, as it were, which are linked with all the others, which however contradict all the other sites" (15). But again, in the framework of this project, Foucault's focus on space at the expense of place, limits the full application of this model.

53. The camp near Biloxi has been referred to as a "meme" camp and has spawned other relief efforts by Burning Man participants.

CHAPTER 10. FROM MANILA VILLAGE TO NEW ORLEANS

1. The following excerpt gives the context for this chapter's epigraph: "[T]he well-grieved Achaeans heaped the corpses on the pyre, inwardly grieving, and when they had burned them with fire they went to the hollow ships. . . . And in it they made gates, close-fastening, so that through them there might be a way for the driving chariots. And outside it they dug a deep trench hard by, wide and great, and they planted stakes in it." My use of "city of ships" comes from Indra Kagis McEwen, *Socrates' Ancestor: An Essay in Architectural Beginnings* (Cambridge: MIT Press, 1993), 79–122. McEwen writes, "it is clear that the fleet of Achaean ships beached at the mouth of the Maeander during the ten-year siege of Troy was thought of as a city" (93).

2. See C. C. Lockwood, *Atchafalaya: America's Largest River Basin Swamp* (Baton Rouge:

Beauregard Press, 1981), for a series of photographs that document a family living on a fixed barge that has been converted to a houseboat.

3. See Benjamin D. Maygarden and Jill-Karen Yakubik's *Bayou Chene: The Life Story of an Atchafalaya Basin Community* (New Orleans: U.S. Army Corps of Engineers, 1999). The full text can be found at http://www.mvn.usace.army.mil/prj/hist_arch/bayou_chene/bayou_chene.pdf. The flood of 1927 led many of the settlers to move away from the area, although some built scaffolding inside their homes to adjust to the new water level while others transferred their possessions to houseboats or barges.

4. Patricia O'Grady, "Thales of Miletus," Internet Encyclopedia of Philosophy, http://www.utm.edu/research/iep/t/thales.htm.

5. For a similar phenomenon in north Florida and southern Georgia, see also Stetson Kennedy, "Trembling Earth," in *Palmetto Country* (Tallahassee: Florida A&M University Press, 1989), 17–23. This section portrays life in the Okefenokee Swamp, a name derived from an Indian term meaning "trembling earth." Kennedy also describes the islands, known as "trembling earth houses," one of which, "Billy's Island," was the famous encampment of Billy Bowlegs.

6. Rodrigo Perez, "Philippine Architecture," *AIA Journal* 34 (October 1960): 40–46.

7. Rodrigo Perez, "Tausug," in *Encyclopedia of Vernacular Architecture of the World* (Cambridge: Cambridge University Press, 1997), 1206.

8. For more detailed histories of this immigration, see Marina E. Espina, *Filipinos in Louisiana* (New Orleans: Laborde, 1988), and Betsy Swanson, *Historic Jefferson Parish from Shore to Shore* (New Orleans: Pelican, n.d.).

9. Aaron L. Shalowitz and Michael W. Reed, "United Nations Conferences on the Law of the Sea," in *Shore and Sea Boundaries* (Washington, D.C.: Office of Coast Survey, National Oceanic and Atmospheric Administration, 1962) vol. 1, pt. 3, chap. 2, 211. Note: The three volumes have also been published online at: http://chartmaker.ncd.noaa.gov/hsd/shallow.htm.

10. *Louisiana Boundary Case*, 394 U.S. 66 (1969), in Shalowitz and Reed, "The Tidelands Litigation," in *Shore and Sea Boundaries*, vol. 3, pt. 1, 64. (Washington, D.C.: Office of Coast Survey, National Oceanic and Atmospheric Administration, 2000).

11. Ibid., 394 U.S. 11, 32 (1969), in Shalowitz and Reed, "The Tidelands Litigation," in *Shore and Sea Boundaries*, vol. 3, pt. 1, 92.

12. Aaron L. Shalowitz and Michael W. Reed, "The Tidelands Litigation," in *Shore and Sea Boundaries*, vol. 3, pt. 1, 72.

13. Ibid., 64.

14. Primarily associated with the Louisiana Delta region (though appearing elsewhere), mudlumps are cones of clay that form in the intermediate coastal area between the tidal forces of the Gulf of Mexico and the Mississippi River's current. In the mudlump's formation, the weight of the river's sedimentary deposits pushes down on the plastic clay stratifications along the coast. As the mudlumps begin to form through a process of extrusion, the tides reinforce the uplift of the clay, and eventually the cones reach the surface of the water. With the possibility of rapid formation and growth (as much as four feet per day), the mudlumps

are a dynamic element that changes form and position. Geologically, they are understood as fields of sediment that after formation are primarily moved around by tidal push and pull and the river's current. The mudlumps frequently serve as matrices for fossils and shipwrecks as a result of their formation from a natural dredging of the ocean floor by the tides.

15. Shalowitz and Reed, "The Tidelands Litigation," in *Shore and Sea Boundaries*, 72.

16. Bricolage, in this case, is a kind of making not just through orality but also through the active rituals, which for the residents of Manila Village is the "dancing of the shrimp" discussed in the next section. Overall, the extended series of excerpts from "The Tidelands Litigation" can be read as what Michel de Certeau has termed the "travel literature" of juridical discourse: "These 'operations of marking out boundaries,' consisting in narrative contracts and compilations of stories, are composed of fragments drawn from earlier stories and fitted together in makeshift fashion (*bricolés*). In this sense, they shed light on the formation of myths, since they also have the function of founding and articulating spaces. Preserved in the court records, they constitute an immense travel literature, that is, a literature concerned with actions of organizing more or less extensive social cultural areas. But this literature itself represents only a tiny part (the part that is written about disputed points) of the oral narration that interminably labors to compose spaces, to verify, collate, and displace their frontiers." See Michel de Certeau, "Spatial Stories," in *The Practice of Everyday Life*, trans. Steven Rendall (Berkeley and Los Angeles: University of California Press, 1988), 122–23.

17. Ibid., 122.

18. In 1965, Hurricane Betsy destroyed Manila Village. Today, only the pilings remain.

19. H. R. Padgett, "The Marine Shell Fisheries of Louisiana," Ph.D. diss., Louisiana State University, 1960.

20. S. Jeffress Williams, Shea Penland, and Asbury H. Sallenger, eds., *Atlas of Shoreline Changes in Louisiana from 1853 to 1989* (n.p.: U.S. Geological Survey and the Louisiana Geological Survey, 1992).

21. It has been noted that these pots were "literally big enough for a horse to swim in." Jim Bradshaw, "South Louisiana Industry Once Danced to a Different Tune," *Lafayette Daily Advertiser*, July 29, 1997.

22. "History of Chinese Shrimping" (Galveston, Tex.: Southeast Fisheries Center), http://galveston.ssp.nmfs.gov/shrimpfishery/history.htm.

23. Espina, *Filipinos in Louisiana*, 38–39, 50; Marina Espina at the "Conference on Filipinos in America," Pasig City, Philippines, August 5–7, 1998.

24. William Dale Reeves, *Westwego: From Cheniere to Canal* (Westwego, La.: Daniel P. Alario, 1996).

25. Ibid. Just as the vernacular houseboat constructions were developed into the platform dwellings and synthesized with the drying platforms, the traditionally woven Chinese shrimp baskets are related to the standards of measure utilized in the West for measurement and trade. Techniques of construction are maintained while the form is allowed to vary based on local economic standards and environmental conditions.

26. Charles Sanders Peirce, *The Collected Papers*, ed. Charles Hartshorne and Paul Weiss (Cambridge: Belknap Press of Harvard University Press, 1960), 4: 413–14, par. 531, quoted in

Anthony Vidler, "Diagrams of Utopia," *Daidalos* 74 (2000): 6. Deriving its conception from his thoughts on icon, Peirce defines diagram as "a *representamen* which is predominantly an icon of relations and is aided to be so by conventions" (4: 341, par. 418). The *representamen* can stand in place of something else while retaining its mental effect. In what Peirce calls the "triadic relation" of subject, object, and "interpretant," the *representamen* is the subject. In terms of representation, the *representamen* (similar to a sign) *mediates* between an object and an interpreting thought. Ultimately, a diagram *represents* relations.

27. Peter Eisenman, "Diagram: An Original Scene of Writing," in *Diagram Diaries*, by Eisenman (New York: Universe, 1999), 28.

28. Eisenman expands on this idea of diagrammatic function: "The diagram then is both form and matter, the visible and the articulable. Diagrams for Deleuze do not attempt to bridge the gap between these pairs, but rather to attempt to widen it, to open the gap to other unformed matters and functions, which will become formed. Diagrams, then, form visible matter and formalize articulable functions." Ibid., 30.

29. While the concept of the diagram as the generator for architectural process is important for this discussion, R. E. Somol's opposition of a diagrammatic practice (such as Eisenman's) to a tectonic vision of architecture limits the dual role of the diagram, specifically in the study of a phenomenon such as Manila Village's platforms. See R. E. Somol, "Dummy Text, or the Diagrammatic Basis of Contemporary Architecture," in *Diagram Diaries*, by Peter Eisenman (New York: Universe, 1999), 24.

30. Giambattista Vico, *The New Science* (Ithaca, N.Y.: Cornell University Press, 1984), 1.92.33.

31. Within an inherent multiplicity mixed with identity, the platforms of Barataria Bay serve as sites for the study of place, territory, and itinerancy.

32. Michel Serres continues, "Never with the stations from which it comes, to which it goes, and by which it passes. Never to the things as such and, undoubtedly, never to subjects as such. Or rather, to those points as operators, as sources of relations. And that is the meaning of the prefix para- in the word parasite: it is on the side, next to, shifted; it is not on the thing, but on its relation. It has relations, as they say, and makes a system of them. It is always mediate and never immediate. It has a relation to the relation, a tie to the tie." *The Parasite*, trans. Lawrence R. Schehr (Baltimore: Johns Hopkins University Press, 1982), 38–39.

33. Going back to its connotative meaning, this parasitism is not detrimental but often beneficent.

34. Michel de Certeau, "'Making Do': Uses and Tactics," in *The Practice of Everyday Life*, 37.

35. Ibid., 41. A comparison might also be made to Bruno Latour's archipelago model within his critique of actor-network theory.

36. De Certeau also relates such a protean construct to the city postulated by Kandinsky in *Du spirituel dans l'art* (1969).

37. De Certeau, *The Practice of Everyday Life*, 110.

38. Gerald Adkins, "Shrimp with a Chinese Flavor," *Louisiana Conservationist* 25, no.7–8 (1996): 20–25.

39. See also chapter 1 for the discussion of the *mattang* maps and navigation. Michel Serres describes this process as well as the resulting souplike situation: "The chain is not a chain of chance either, it would remain meticulously broken. It is the chain of contingency; the recruiting takes place through tangency, by local pulls and by degrees, by word of mouth, from one mouth to the other. It emerges from the sea noise, the nautical noise, the prebiotic soup." See *Genesis*, trans. Geneviève James and James Nielson (Ann Arbor: University of Michigan Press, 1995), 71–72.

40. Alan Colquhoun has pointed out that regionalism is an internalized question of individualism and the nation-state. See "The Concept of Regionalism," in *Postcolonial Space(s)*, ed. G. B. Nalbantoglu and C. T. Wong (New York: Princeton Architectural Press, 1997), 13–23.

41. See Fredric Jameson, *Geopolitical Aesthetic: Cinema and Space in the World System* (Bloomington: Indiana University Press, 1995), 11.

42. Vidler, "Diagrams of Utopia."

43. See also chapter 5 for an extended discussion of Michel Foucault, "Other Spaces," *Lotus International*, nos. 48–49 (1986). Related to the notion of camp, an example used to describe the classification of "imaginative heterotopia" is the colony, which for him includes the "placeless place" of the ship. Also, see Vittorio Gregotti, "On Atopia," in *Inside Architecture*, trans. Peter Wong (Cambridge: MIT Press, 1996), 75–82. The space of platform architecture includes qualities of atopia, heterotopia, and, as noted previously, utopia. See the previous discussion on the subject in the "Siting Camp" section of chapter 5.

44. See note 29 in this chapter.

45. The work of John Ruskin, whose acts of digression in his *Brantwood Diaries* and other works are similar to the idea of the tangent introduced previously, relates to the builder/bricoleur: "I work down or up to my mark, and let the reader see process and progress, not caring to conceal them, but this book [*Proserpina*] will be nothing but process." *The Works of John Ruskin*. ed. E. T. Cook and Alexander Wedderburn, Library Edition (London: 1903–12), 25: 216.

46. Compare to the additional component of C. S. Peirce's diagram, which for him becomes "pure dream." *The Collected Papers*, 3: 211, par. 362. See the discussion of Peirce in chapter 5.

47. "There is nothing left of Minilla [*sic*] Village, only a few pilings. E. J." E-mail to the author from Captain E. J. Plaisance, October 1, 2006. See also note 18.

48. As Anthony Vidler concludes his essay "Diagrams of Utopia": "the diagram has come to signify . . . the instrumentalization of a world to be made by social, political, and intellectual endeavor. The diagram in this context can act to galvanize the discourse only if both political form and architectural form are entered into its equation" (13).

49. In late August 2006, Rockey Vaccarella traveled to Washington D.C., in a "mock" FEMA trailer to call attention to the uncertain status of New Orleans residents.

50. Robert Tannen, an urban planner and artist, has proposed the use of an elevated grid to raise FEMA trailers above the flood-prone Seventh and Ninth Wards. See http://www.ogdenmuseum.org/rebuilding_new_orleans.

51. Groups like the CITYbuild Consortium of Schools and Project Locus have provided

rebuilding efforts with students, faculty, and community members and have also sought to address challenges of urban environments that preceded Hurricane Katrina. See http://www.citybuild.org/and http://www.projectlocus.org. The magazine *Architectural Record*, along with the Tulane School of Architecture, sponsored a competition in June 2006 with two parts: "High Density on the Ground" and "New Orleans Prototype House." And the Tulane School of Architecture has also developed the URBANbuild project, which will construct many of the winning projects.

52. Refer to Allan Wallis's discussion of Detroit's municipal camps in *Wheel Estate: The Rise and Decline of Mobile Homes* (New York: Oxford University Press, 1991), 71–76.

53. Postdisaster camps reframe categories of statistical tabulation. When St. Bernard Parish released its postdisaster demographic, we read of about 200 "residents in homes" (out of 8,800 residents); 2,000 government workers in trailers or homes; 2,500 trailer residents with families; and a daily influx of workers and disaster officials, doubling the base population. In Florida, FEMA trailer occupancy peaked at 17,075 households after Hurricane Charley.

54. Understanding the "migrational city" as a "metaphorical city" spliced within the "clear text of the planned and readable city," Michel de Certeau rereads Kandinsky's *Concerning the Spiritual in Art* through the exercise of painting in particular: "The childhood experience that determines spatial practices later develops its effects, proliferates, floods private and public spaces, undoes unreadable surfaces, and creates within the planned city a 'metaphorical' or mobile city, like the one Kandinsky dreamed of: 'a great city built according to all the rules of architecture and then suddenly shaken by a force that defies all calculation.'" *The Practice of Everyday Life*, 110.

55. The urban image of Manhattan had already changed significantly from de Certeau's syntactic city before September 11, 2001, and subsequent changes after the destruction of the World Trade Center are well documented, particularly in the competitions and discussions of its redevelopment.

56. See Foucault, "Of Other Spaces," 15.

57. In another permutation of the heterotopia, bus terminals became camps of detainment. Camp Greyhound in New Orleans' Union Passenger Terminal held fugitives and violators of the postdisaster curfew.

58. October 2006 estimates showed more than 1,600 residents of Renaissance Village.

59. On April 6, 2006, 8,000 trailers were promised by June. As many as 1,500 trailers were in place within "enclosed" urban sites, 10,000 trailers were in yards and side lots, and 10,000 residents (or "refugees") were still waiting for trailers to be delivered.

60. Security officers of Renaissance Village refer to the camp as the "installation." In this respect, the postdisaster camp risks militarization.

61. Hurricane Charley devastated areas of Charlotte County and Punta Gorda in August 2004, and residents remained in FEMA trailers through November 2006.

62. David Greene writes: "instant villages . . . (camping scene not included)." "Gardener's Notebook," in Archigram, ed. Peter Cook (New York: Princeton Architectural Press, 1999), 110–15.

63. Some camps combine house and trailer. This coupling reverses the romantic notion

of living on the construction site. Marcel Breuer designed around his client's Spartan trailer (Mansion Model) for the Wolfson Trailer House.

64. See the discussion of slabs at Slab City in chapter 9.

65. The 1974 Stafford Act prevents FEMA's provision of permanent residential construction. FEMA contends that the postdisaster dwelling unit itself cannot be a house, or a home. A diverse set of vehicles and units provide shelter in camps. FEMA Form 81–96, a worksheet for documenting loss after a disaster, registers this array: "Manufactured (Mobile) Home/ Travel Trailer." The protean camping unit produces a new syntax: the interpolating parenthetical and the hybridizing slash-mark. The diagram of the housing unit in 81–96 becomes an additional formulary for a lost domestic life, its impossibly elongated blank space a memory theater of displacement. FEMA officials rejected the "Katrina Cottage" in its mission to provide temporary housing assistance, not houses.

66. The community plaza, with a children's play area, was funded by actress Rosie O'Donnell's "For All Kids Foundation" and includes three playgrounds, an Early Head Start and Head Start programs, a computer lab, and a counseling center, among other services. One of the stated goals of the foundation was to provide a place that would allow a "sense of community" to develop among Renaissance Village's residents.

67. FEMA also proposed areas of the city's university campuses—reminiscent of post–World War II campuses transformed by the reconstruction of military camp buildings to accommodate the returning student-soldiers.

68. His order was filed on March 24, 2006. Mayor Nagin feared conflict with the established neighborhood's residents. (NIMBY is an acronym for "not in my backyard.")

69. FEMA has noted that its postdisaster camps can be demobilized with respect to "local concerns."

70. A risk with these types of urban infill is that they become enclaves, lacking any integration within the urban context.

71. See the discussion in chapter 6.

CHAPTER 11. BREAKING CAMP

1. With camps, paradoxical conditions arise such as mobile fixity, unstable permanence, and chronic itinerancy. Michel Butor captures this idea in his description of the unsolidified permanence and hardened ephemerality of the vernacular camp constructions that became Istanbul: "the encampment that has settled, but without solidifying completely." See the discussion in chapter 4 of how the vernacular is made. The encampment, though "enlarged" and "improved," does not lose its transient quality, even within a growing permanency.

2. In 2002, the town of Bridgeville sold for $1.8 million to Bruce Krall. In 2006, the town again went up for sale on eBay and was purchased by Daniel La Paille in August for $1.25 million. La Paille plans to refurbish the community's infrastructure and to strengthen its identity.

3. John Leland, "Trying to Stay Put in Florida Mobile Homes," *New York Times*, June 22, 2003.

4. In 2006, the 154-unit America Outdoors RV Park was also converted into the 90-unit Playa Cristal Resort, with unit prices ranging from $850,000 to $3 million.

5. This section seeks to address one of the concluding questions of whether the exodical nature of nomadic hesitation can bring us back to place. Or, if modeled and understood as campsite, does the hesitation present us with new ways of thinking and making place?

6. The consequences of this nomadic hesitation are found in the uneven development of Hotel Palenque as a place, Old Havana as a city, and *gecekondu* as a legal framework. Hotel Palenque exemplifies the concurrence of breaking and siting, dismantling and constructing. In his documentation of the hotel, Robert Smithson notes: "You can see that instead of just tearing it all down at once, they tear it down partially so you're not deprived of the complete wreckage situation. It's not often that you see buildings being both ripped down and built up at the same time." Such modification is registered as changes of degree rather than of kind. Smithson quoted in Robert A. Sobieszek, *Robert Smithson: Photo Works* (Los Angeles: Los Angeles County Museum of Art, 1993), 116–17. For the most part dating from the Spanish colonial period, the urban apartments, or *ciudadellos*, of Habana Vieja are divided up by *artesanos del espacio*, who provide housing for growing families by magically making space out of nothing. This term translates literally as "artisans of space," but it identifies those who are known in Cuba as "magical carpenters." The band Los Van Van, formed in 1969, sings the story of the builders in a popular Cuban song, "Artesanos del Espacio." As acts of resistance, the partitions and the spaces that the *artesanos del espacio* create serve as urban campsites for a transient, rural population. Out of necessity subverting the material permanence of the colonial-era stone buildings, the *ciudadellos* combine construction and deconstruction (or partial demolition) in the making of temporary spaces for living. In Ankara, Turkey, legal codification provides a stable, though invisible, infrastructure for the construction of temporary dwellings. By law, squatters are allowed to maintain occupation of a site if a suitable structure is built during the period of one night. While not a strict inversion of stability, Ankara's *gecekondu*, literally "house built in one night," does reflect another relevant problematic—a permanence that requires speed for its realization. Mary-Ann Ray, "Gecekondu," in *Architecture of the Everyday*, ed. Steven Harris and Deborah Berke (New York: Princeton Architectural Press, 1997), 153–65. What implications do these reversals and coincidences of the temporary and the permanent have for the architecture of dwelling?

7. See John Brinckerhoff Jackson, "The Movable Dwelling," in *Discovering the Vernacular Landscape* (New Haven: Yale University Press, 1984).

8. The series of aphorisms that follow in this chapter explores Nietzsche's grounding without ground and provides starting points for places of hesitation.

9. G. P. Putnam's Sons published Jessup's book in 1921.

10. Judith Scheine, *R. M. Schindler* (New York: Phaidon, 2001), 33.

11. Ibid.

12. Scheine notes that the building permit issued by the officials in Los Angeles was "revocable at any time." Ibid., 111.

13. R. M. Schindler, "A Cooperative Dwelling," *T-Square* 2 (February 1932): 21, in August

Sarnitz, *R. M. Schindler Architect: A Pupil of Otto Wagner between International Style and Space Architecture* (New York: Rizzoli, 1988), 49.

14. In the opening line of his essay "Modern Architecture: A Program," Schindler writes: "The cave was the original dwelling. . . . To build meant to gather and mass material, allowing it to form empty cells for human shelter." Scheine, *R. M. Schindler,* 80.

15. Having studied under Otto Wagner at the Imperial Technical College of Vienna, Schindler would have been very familiar with Semper, particularly through Wagner's design to complete Semper's 1869–70 plan for an Imperial Forum with the Hofburg Throne Room. Schindler would also have known Semper through Adolf Loos, with whom he associated and who critiqued the German architect's concept of style (*Der Stil in der technischen und tektonischen Kunsten*). Contemporary critics such as Mark Wigley have realigned Loos and Semper through their understanding of surface and the "clothing" of architecture. See also Andrew Benjamin's *Architectural Philosophy* (2000).

16. Schindler's A. E. Rose Beach Colony project dates from 1937. See "Clearing for Play" in chapter 3.

17. Deleuze and Guattari write: "The nomad is there, on the land, wherever there forms a smooth space that gnaws, and tends to grow, in all directions. The nomad inhabits these places, he remains in them, and he himself makes them grow, for it has been established that the nomad makes the desert no less than he is made by it. He is a vector of deterritorialization. He adds desert to desert, steppe to steppe, by a series of local operations the orientation and direction of which endlessly vary" (*Nomadology,* 53).

18. Schindler, "Modern Architecture: A Program" (Vienna 1912), in Scheine, *R. M. Schindler,* 81. Architect Eileen Gray addressed the "camping method" in her house designs as a necessary way of life, but one that she hoped would pass as wartime effects and the freneticism of social life receded. For her, it was a necessity that the camping life remained a temporary concept. But for Schindler, modern dwelling must not simply freeze the temporary whims of the inhabitant. Gray writes, in the winter of 1929: "The camping style is only a temporary concept, and the creations that it inspires are unquestionably precarious. By eliminating all intimacy, it leads to an impoverishment of the inner life. The truly civilized man requires a certain formal elegance: he knows the propriety of certain gestures; he needs to be able to isolate himself." Eileen Gray quoted in Caroline Constant's translation of the essay "Maison en bord de mer," by Gray and Jean Badovici, in appendix 4 of *Eileen Gray* (London: Phaidon, 2000), 241.

19. Note that Cage recalls staying at the Kings Road house for much longer; in Thomas Hines's "Cage Was Not Yet Cage," he recalls his stay as "probably a little less than a year." In Elizabeth A. T. Smith and Michael Darling, *The Architecture of R. M. Schindler* (Los Angeles: Museum of Contemporary Art, 2001), 110. Smith and Darling note that Cage returned in April 1935 to host a concert of classical shakuhachi music.

20. Cage to Pauline Gibling Schindler, April 15, 1935, Collection 980027, Getty Research Institute for the History of Art and the Humanities, Los Angeles, in Smith and Darling, *The Architecture of R. M. Schindler,* 110.

21. Michel Serres, *The Natural Contract,* translated by Elizabeth MacArthur and William Paulson (Ann Arbor: University of Michigan Press, 1995), 99–100.

22. Massimo Cacciari (along with) notes the related idea of desertion in Heidegger's later work ("Eupalinos or Architecture," *Oppositions* no. 21 [1980]: 106–16). In subsequent commentary, the architect Ignasi de Solà-Morales salvages Cacciari's nihilism (in which desertion replaces dwelling) through the idea of an archaeology of weak architecture. *Differences: Topographies of Contemporary Architecture* (Cambridge: MIT Press, 1997).

23. This refastening can also be understood as a reification that works back from place to camp. In this sense, camp reifies the abstractions and ideas of place in its mental and often physical reconstruction of place. Camp becomes one of many possible materializations of place. And that which "casts off" can be read as a detail temporarily detached from or chronically lodged within a territory. Already complicating the distinction of temporary and permanent, the vessels of camping also narrate relations between detail and territory. "They [nomads of the sea] do not grasp an itinerary as a whole, but in a fragmentary manner, from campsite to campsite." José Emperaire, *Les nomades de la mer* (Paris: Gallimard, 1954), 225, quoted in Gilles Deleuze and Felix Guattari, *Nomadology: The War Machine*, trans. Brian Massumi (New York: Semiotext[e]: 1986), 133.

CHAPTER 12. DEPARTING CAMP

1. For Thoreau, the place of sitting is the place of living: "Wherever I sat, there I might live, and the landscape radiated from me accordingly. What is a house but a *sedes*, a seat?" *Walden* (Princeton: Princeton University Press, 2004), 81.

2. Thoreau also writes: "Well, there I might live, I said; and there I did live, for an hour, a summer and a winter life; saw how I could let the years run off. . . . The future inhabitants of this region, . . . may be sure that they have been anticipated." Ibid.

3. The detail thus serves as the locus of this paradox of proximity and distance—which leads back to the discussion of Bergson's space and place at the conclusion of the section on Gibsonton. In his treatment of Aristotle's ideas of place, Bergson wrote: "A body possesses a place on the condition of being at a remove from this place." "L'idee de lieu chez Aristotle," *Les Etudes Bergsoniennes* 2 (1949): 80. "This "being at a remove" can occur both in potentiality (such as in the mobility of the recreational vehicle, trailer, or camper) and in materiality or technique (such as the imported platform construction and shrimp-drying processes in Manila Village). Also, though alluding to his understanding of place as position, the contradictory statement combining possession of a place and a requisite distance does hint at the possibility of being there and here simultaneously.

4. The constructs of Burning Man and Slab City suggest camping as an analog for practicing and making place within the Internet's electronic landscape. See chapter 9.

5. In *The Meaning of Truth*, James writes, "My thesis is that the knowing here is *made* by the ambulation through the intervening experiences" (Cambridge: Harvard University Press, 1975). The ambulation of James's thesis points to a peripatetic practice that does not rely on movement from one point to another destination; rather the "walker" moves with thought. The "intervening experiences" then occur not within a network but in a fluid matrix of thinking and making. This configuration is reminiscent of Michel Serres's reading of Jules Michelet's *Sea* as the soup, which is "milk, blood, a solution of mineral salts, [and] an electri-

cal flux." *Hermes: Literature, Science, and Philosophy,* ed. Josué V. Harari and David F. Bell (Baltimore: Johns Hopkins University Press, 1982), 35. The soup synthesizes hylozoism (animation of matter), vitalism, and encyclopedism into a reservoir, or thermodynamic model.

6. I have derived this phrase from Casey's discussion of place-as-pragmatic. *The Fate of Place* (Berkeley and Los Angeles: University of California Press, 1997), 246. Place-as-pragmatic combines with duration in the zone where things are worked on.

7. Sometimes these overturned vessels are used for shelter within the camp.

8. See John Reps's account of the evolution of the Laws of the Indies, begun by Philip II as a set of rules governing the layout and establishment of Spanish colonies. Reps contends that the Spanish authorities would have known of Machiavelli's *Arte della guerra* (1521), which includes detailed plans of "A Fortified Camp and Lodgings" (see figures 6 and 7 in Machiavelli's work). In addition to Roman castramentation theories published in the mid-sixteenth century, authorities provided plans for civil communities based on the Roman *castra* and related military settlements. *Town Planning in Frontier America* (Princeton: Princeton University Press, 1969), 46. See also Reps on the rules governing the layout and function of mining camps. *Cities of the American West: A History of Frontier Urban Planning* (Princeton: Princeton University Press, 1980), x, 195–237, and 491–522.

9. The Deleuzian concept of the relative global offers the condition diametrically opposed to the local absolute. The pairing of these terms occurs within a "field of becoming" where the "local becomes the absolute" and the "absolute becomes the local"—crossings of these movements may occur at any number of points within this hypothetical "field," particularly in recalling Casey's contention that "place is everywhere."

10. Less directly, these platforms can be related to the Achaeans' "floating city" of ships and the concomitant suggestion that a camp can be an interconnected system of heterotopic units floating on water's fluid and inherently mobile matrix.

11. The story is told that the rationale behind the Indians' repeated attacks on the original Braden Castle was to reclaim their access to these sacred places.

12. In the tradition of municipal camps and the siting of Braden Castle, the camps are tactical locations next to, over, around, and *in* historically significant sites. Braden Castle Park also provides a counterpoint to New Urbanist constructions such as Seaside, Florida, which ultimately lend themselves to the regulated place of leisure that the park has always been. Robert Davis, Andres Duany, and Elizabeth Plater-Zyberk reviewed what they considered real towns in their formulation of the urban and architectural code for the new town. It is perhaps ironic then that the town they did not review is closer to what they developed than the "real" Panhandle towns of De Funiak Springs, Florida, and Eufaula, Alabama.

13. Gilles Deleuze, *Bergsonism,* trans. Hugh Tomlinson and Barbara Habberjam (New York: Zone, 1991), 32.

14. Bergson writes: "Our concept of duration was really the translation of a direct and immediate experience." *Duration and Simultaneity: Bergson and the Einsteinian Universe,* trans. Leon Jacobson (Manchester: Clinamen, 1999), xxvii.

15. Deleuze, *Bergsonism,* 49. Deleuze offers here a reading of Bergson's concept of duration. This "new space" also relates to de Solà-Morales's "place-as-event" and Casey's "place-as-pragmatic."

16. Thoreau later writes in *Walden:* "Time is but the stream I go fishing in" (98).

17. In terminology that also applies to the place of camps, Michel Serres characterizes this threshold as crossing, *chi*, chimera, and *chora*. Camps also can occur at literal and phenomenal crossroads, can be constructed as bricolaged assemblages (both territorially and in terms of detail), and can be understood as matrices in which notions of home and dwelling are suspended.

18. Serres notes, "she makes and undoes this cloth that mimes the progress and delays of the navigator, . . . on board his ship, the shuttle that weaves and interweaves fibers separated by the void, spatial varieties bordered by crevices." *Hermes*, 49.

19. Serres's summation of the broken threshold could also describe the method of camping: "This is a discourse that weaves a complex, in the first sense of the term, that connects a network, that traces a graph upon space." Ibid., 47.

20. To continue with Serres's discussion: "a loop that turns back upon itself toward a previous crossroads and strongly reconnects the spatial complex. I began with *local singularity* of space, and I finish with a *global law* that is invariably written as the connection of what is separated." Ibid., 48; italics added.

21. In addition to the etymological derivation discussed in the second chapter, words related to the vernacular affirm this connection to home: "indigenous" and "autochthonous." As noted by both J. B. Jackson and Dante, the strength of the vernacular is its ability to take foreign materials, techniques, and ideas and transform them to make a place, often a new place, a home. Equating the vernacular to his own exiled condition, Dante himself inhabits the vernacular language as a linguistic home away from home. Jackson speaks of the basis of the American understanding of home being one of a mobile indigenous architecture. As sites for research and practice, camps provide a laboratory for investigating all of the workings of the vernacular. As sets of relations, the camps studied clearly address this paradox of the internal incorporation of the external, the foreign, and the strange.

22. As a living organism, the vernacular provides a working site for these constructions of place. The vernacular becomes an environment for both William McDonough's Hannover Principles within the "living world" of the nature-building interdependence and Bergson's "lived act" of intuition. The vernacular provides a site for two "contracts" to be drawn up: the natural contract between built and natural environment and the more ontological (and for Bergson intuitive) contract between true experience and the environment of differences of kind that Bergson finds in duration. Camps and the ways of practicing and thinking they suggest afford vernacular sites for the review of these contracts.

23. See the discussions of home with reference to David Morley in chapter 1 and to Gaston Bachelard in chapter 5.

24. Camp thus navigates what John Hejduk has called place-actual and place-imagined. *Riga* (Berlin: Aedes, 1988), n.p.

25. Excerpted in William Howarth, ed., *Walking with Thoreau* (Boston: Beacon, 2001), 29. Note that William Wordsworth often camped in the North Lake District of England.

26. In the mid-nineteenth century, these camps were also part of an image constructed to attract tourists to the remote areas of the Adirondacks; these images were derived from a series of symbols: the guide, the Native American, the hermit, the guide boat, the open

camp, and the pack basket. See Jackie Day, http://www.neh.gov/news/humanities/1999-07/adirondacks.html. See also *Two Adirondack Hamlets in History: Keene and Keene Valley* (Fleischmanns, N.Y.: Purple Mountain Press, 1999), ed. Richard Plunz. Note in particular the chapter titled "Urban Intellectuals in the Valley," in which Plunz tracks the "intellectual discovery" of the Adirondacks from Thomas Cole through Stillman and Emerson to William James at Putnam Camp.

27. Notably absent from the scene is Henry Wadsworth Longfellow, who refused to attend after Emerson announced his intention to carry a gun. See Anne Ehrenkranz's introduction to William James Stillman, *Poetic Localities* (New York: Aperture, 1988): 16–17. See also William James Stillman's essay "The Philosophers' Camp," in *The Old Rome and the New and Other Studies* (1898; reprint, Freeport, N.Y.: Books for Libraries Press, 1972), 272–96.

28. Ralph Waldo Emerson, "Adirondacs"—the poem written following the experience at Camp Maple in August 1858.

29. It might be said that many of us go camping to make sure that we have "really *lived*," and in contemporary travel, this deliberate life takes on many forms. Traveling on Interstate 75 south to Tampa, Florida, I remember seeing a bumper sticker on the back of a Winnebago recreational vehicle: "Living the good life 30' at a time."

30. William James, "What Pragmatism Means," in *Pragmatism* (New York: Meridian Books, 1955), 41. For an exhaustive treatment of Putnam Camp's histories, see George Prochnik's *Putnam Camp: Sigmund Freud, James Jackson Putnam, and the Purpose of American Psychology* (New York: Other Press, 2006).

31. James, "What Pragmatism Means," 42.

32. Accordingly, the places of camp are outlined by the practicality of the process itself. Camping pragmatism is not, however, a reductive procedure emptied of complexity of meaning or significance. Instead the practice of place through camping results in a synthesis of both the vagaries and the poetics of existence. Practicalities here are not solely about efficiency but relate by necessity to the experience of a place.

33. In a letter to his family back in Vienna, Freud wrote: "Of everything I have experienced in America this here is probably the strangest: a camp, you must imagine, in a wilderness in the woods, situated like a mountain meadow. . . . Stones, moss, groups of trees, uneven ground which on three sides merges into densely wooded hills." Sigmund Freud to his family, dated Putnam's Camp, September 16, 1909, translation of a typescript, Adirondack Museum, Blue Mountain Lake, N.Y., cited in *Two Adirondack Hamlets in History: Keene and Keene Valley*, ed. Richard Plunz, 204.

34. Thoreau begins his 1858 journal with the January 1 entry: "I have lately been surveying the Walden woods so extensively and minutely that I now see it mapped in my mind's eye—as, indeed, on paper—as so many men's wood-lots, and am aware when I walk there that I am at a given moment passing from such a one's wood-lot to such another's." In Raymond R. Borst, *The Thoreau Log: A Documentary Life of Henry David Thoreau 1817–1862* (New York: G. K. Hall, 1992), 466.

35. See August 23 entry in ibid., 492–93.

36. Emerson provided lodging to Thoreau periodically during the 1840s and 1850s.

37. Camping practice is proposed as an analog for an interpretive method of studying and operating within places and as a device of demonstrative invention. Camping allows for a flexible negotiation of place—such that space becomes the practiced place of Michel de Certeau. Such method combines John Ruskin's "caravannish manner" and his "patchwork mentality." By moving in a digressive path from personal reflection to observation to Ruskinian verity, moving thus from camp(site) to campsite, Ruskin outlines a practice of pure process that can be traced and read as a loosely drawn itinerary that is followed to varying degrees. Such is a peripatetic practice. In camping, learning occurs in moving. The Peripatetic school practiced this notion by moving within the *campus* of the lyceum. Within this "open field," the Peripatetics walked from place to place under the maxim *solvitur ambulando*—the solution is in the walking.

BIBLIOGRAPHY

Adkins, Gerald. "Shrimp with a Chinese Flavor." *Louisiana Conservationist* 25, no. 7–8 (1996): 20–25.

Agee, James, and Walker Evans. *Let Us Now Praise Famous Men.* Boston: Houghton Mifflin, 1960.

Alberti, Leone Battista. *Dinner Pieces.* Translated by David Marsh. Binghamton, N.Y.: Center for Medieval and Early Renaissance Studies, 1987.

Alexander, Christopher. *The Production of Houses.* New York: Oxford University Press, 1985.

Alighieri, Dante. *De vulgari eloquentia.* Edited and translated by Steven Botterill. New York: Cambridge University Press, 1996.

Andreotti, Libero, and Xavier Costa, eds. *Theory of the Dérive and Other Situationist Writings on the City.* Barcelona: ACTAR, 1994.

Aristotle. *Metaphysics.* Translated by Hugh Tredennick. Cambridge: Harvard University Press, 1936. Bk. 1, sec. 3, lines 1–14, pp. 17–25.

Assad, Maria L. *Reading with Michel Serres: An Encounter with Time.* Albany: State University of New York Press, 1999.

Auden, W. H. "Prologue: The Birth of Architecture." In *Complete Works of W. H. Auden,* edited by Edward Mendelson. Princeton: Princeton University Press, 1993.

Bachelard, Gaston. *The Psychoanalysis of Fire.* Translated by Alan C. M. Ross. London: Routledge and Kegan Paul, 1964.

———. *Water and Dreams.* Translated by Edith R. Farrell. Dallas: Pegasus, 1983.

Bair, Frederick H., Jr. "Applying Land Use Intensity to Public Regulation." *Urban Land* 26, no. 4 (April 1967).

Bandini, Mirella. "An Enormous and Unknown Chemical Reaction: The Experimental Laboratory in Alba." In *On the Passage of a Few People through a Rather*

Brief Moment in Time: The Situationist International 1957–1972, edited by Elisabeth Sussman, 67–71. Cambridge: MIT Press, 1989.

Baudrillard, Jean. *America*. New York: Verso, 1988.

Beecher, Mary Anne. "The Motel in Builders' Literature and Architectural Publications: An Analysis of Design." In *Roadside America: The Automobile in Design and Culture*, edited by Jan Jennings, 115–24. Ames: Iowa State University Press, 1990.

Bell, Gertrude. *The Desert and the Sown*. London: Heinemann, 1907.

———. *The Letters of Gertrude Bell*. New York: Penguin, 1987.

Benjamin, Andrew. "Parasitism in Architecture." In *Ephemeral Structures in the City of Athens: International Competition Program*, 55–62. Athens: Hellenic Cultural Heritage SA, 2002.

Benjamin, Walter. *The Arcades Project*. Translated by Howard Eiland and Kevin McLaughlin. Cambridge: Belknap Press of Harvard University Press, 1999.

Bergson, Henri. *The Creative Mind: An Introduction to Metaphysics*. New York: Citadel, 1992.

. *Duration and Simultaneity: Bergson and the Einsteinian Universe*. Translated by Leon Jacobson. Manchester: Clinamen, 1999.

———. "L'idee de lieu chez Aristotle." *Les Etudes Bergsoniennes* 2 (1949).

———. *Matter and Memory*. Translated by Nancy Margaret Paul and W. Scott Palmer. New York: Zone Books, 1991.

Berke, Deborah. "Thoughts on the Everyday." In *Architecture of the Everyday*, edited by Steven Harris and Berke, 222–26. New York: Princeton Architectural Press, 1997.

Betsky, Aaron. *Building Sex: Men, Women, Architecture and the Construction of Sexuality*. New York: Morrow, 1995.

Bey, Hakim. "The Temporary Autonomous Zone." In *The Writings of Hakim Bey*, edited by Al Billings. The Hermetic Library. http://www.hermetic.com/bey/taz3.html.

Binkley, Loren O., and Pamela Gibson. "Braden Castle Camp and the Camping Tourists of America." April 1984. Typescript in the Eaton Florida History Room, Manatee County Central Library, Bradenton, Fla.

Booth, Mark. *Camp*. London: Quartet Books, 1983.

Borst, Raymond R. *The Thoreau Log: A Documentary Life of Henry David Thoreau 1817–1862*. New York: G. K. Hall, 1992.

Bradenton Chamber of Commerce. "Bradenton Florida Tourist Club." Unpublished pamphlet, 1929–30. Eaton Florida History Room, Manatee County Central Library, Bradenton, Fla.

Bradshaw, Jim. "South Louisiana Industry Once Danced to a Different Tune." *Lafayette Daily Advertiser*, July 29, 1997.

Bunschoten, Raoul. "The Skin of the Earth: A Dissolution in Fifteen Parts." *AA Files*, no. 21 (Spring 1991): 55–59.

Burch-Brown, Carol. *Trailers*. Charlottesville: University Press of Virginia, 1996.

Burkhart, Bryan, and David Hunt, *Airstream: The History of the Land Yacht*. San Francisco: Chronicle, 2000.

Burns, Carol. "A Manufactured Housing Studio: Home/On the Highway." *Journal of Architectural Education* 55, no. 1 (September 2001): 51–57.

Butor, Michel. "Bricolage: An Interview." By Martine Reid and Noah Guynn. *Yale French Studies* 17, no. 10 (January 1994): n. 84.

———. *The Spirit of Mediterranean Places*. Translated by Lydia Davis. Marlboro, Vt.: Marlboro Press, 1986.

Byam, Wally. *Trailer Travel Here and Abroad: The New Way to Adventurous Living*. New York: David McKay, 1960.

Camp, Paul Eugen. "The Attack on Braden Castle: Robert Gamble's Account." Typescript in the Eaton Florida History Room, Manatee County Central Library, Bradenton, Fla.

Camping Tourists of America. "Braden Castle Park: Constitution and By-Laws of Camping Tourists of America." Braden Castle Park, Bradenton, Manatee County, Fla., 1945.

Carpentier, Alejo. "The Marvelous Real in America." In *Magical Realism: Theory, History, Community*, edited by Lois Parkinson Zamora and Wendy B. Faris. Durham, N.C.: Duke University Press, 1995.

Casey, Edward. *The Fate of Place*. Berkeley and Los Angeles: University of California Press, 1997.

———. "Smooth Spaces and Rough-Edged Places: The Hidden History of Place." *Review of Metaphysics* 51, no. 2 (1997): 267–96.

Certeau, Michel de. *The Practice of Everyday Life*. Translated by Steven Rendall. Berkeley and Los Angeles: University of California Press, 1988.

Coker, Jerry. *Improvising Jazz*. New York: Simon and Schuster, 1987.

Collins, A. Frederick. *How to Build a Motor Car Trailer*. Philadelphia: Lippincott, 1936.

Colomina, Beatriz. "The Media House." In *The City of Small Things*, edited by Mechthild Stuhlmacher and Rien Korteknie, 105–17. Rotterdam: Stichting Parasite Foundation, 2000.

Colquhoun, Alan. "The Concept of Regionalism," In *Postcolonial Space(s)*, edited by G. B. Nalbantoglu and C. T. Wong, 13–24. New York: Princeton Architectural Press, 1997.

Constant. See Nieuwenhuys, Constant.

Cook, Peter. "Plug-In." In *Archigram*, edited by Cook, 36–43. New York: Princeton Architectural Press, 1999.

Corbin, Carla I. "The Old/New Theme Park: The American Agricultural Fair." In *Theme Park Landscapes: Antecedents and Variations,* edited by Terence Young and Robert Riley, 183–212. Washington, D.C.: Dumbarton Oaks Research Library and Collection, 2002.

Corsiglia, Betsy, and Mary-Jean Miner. *Unbroken Circles: The Campground of Martha's Vineyard.* Boston: Godine, 2000.

Cowan, James. *A Mapmaker's Dream: The Meditations of Fra Mauro, Cartographer to the Court of Venice.* Boston: Shambhala, 1996.

Cowgill, Donald Olen. *Mobile Homes: A Study of Trailer Life.* Philadelphia: University of Pennsylvania Press, 1941.

Curzon, George Nathaniel. *Persia and the Persian Question.* London: Longmans, Green, 1892.

Davenport, William. "Marshall Islands Navigational Charts." *Imago Mundi* 15 (1960): 22.

De la Croix, Horst. *Military Considerations in City Planning: Fortifications.* New York: Braziller, 1972.

Deering, Ruth. "A 1921–2 Diary of a Trip to Florida." Typescript, Eaton Florida History Room, Manatee County Central Library, Bradenton, Fla.

Deleuze, Gilles. *Bergsonism.* Translated by Hugh Tomlinson and Barbara Habberjam. New York: Zone Books, 1991.

———. *The Fold: Leibniz and the Baroque.* Translated by Tom Conley. Minneapolis: University of Minnesota Press, 1993.

———. *Negotiations.* Translated by Martin Joughin. New York: Columbia University Press, 1995.

Deleuze, Gilles, and Felix Guattari. *Nomadology: The War Machine.* Translated by Brian Massumi. New York: Semiotext(e), 1986.

———. *A Thousand Plateaus.* Translated by Brian Massumi. Minneapolis: University of Minnesota Press, 1987.

Department of Planning and Growth Management, Hillsborough County, Florida. Hillsborough County Comprehensive Plan, Sec. 3.01.02 SB—Show Business Overlay District/Purpose. "Land Development Code." http://www.hillsboroughcounty.org/pgm.

Derrida, Jacques. *De l'hospitalite autor de Jacques Derrida.* Passe du Vent: Genouilleux, 2001.

Deutsche, Rosalyn. *Evictions: Art and Spatial Politics.* Cambridge: MIT Press, 1998.

DiPietro, Monty. "Tadashi Kawamata at Galerie Deux." Tokyo: Assembly Language, April 10, 1999. http://www.assemblylanguage.com/reviews/Kawamata.html.

Dominy, Eric. *Camping.* New York: McKay, 1978.

Douglass, Lillie Bernard. *Cape Town to Cairo.* Caldwell, Idaho: Caxton, 1964.

Duany, Andres, and Elizabeth Plater-Zyberk. "Urban Code—The Town of Seaside."

In *Seaside,* edited by David Mohney and Keller Easterling, 98–99. Princeton: Princeton Architectural Press, 1990.
Eco, Umberto. *The Island of the Day Before.* Translated by William Weaver. New York: Penguin, 1995.
Eisenman, Peter. "Diagram: An Original Scene of Writing." In *Diagram Diaries,* by Eisenman. New York: Universe, 1999.
Emperaire, José. *Les nomades de la mer.* Paris: Gallimard, 1954.
Espina, Marina E. *Filipinos in Louisiana.* New Orleans: Laborde, 1988.
Federal Writers' Project. *Florida: A Guide to the Southernmost State.* New York: Oxford University Press, 1939.
Fellows, Jay. *The Failing Distance: The Autobiographical Impulse in John Ruskin.* Baltimore: Johns Hopkins University Press, 1975.
———. "Janusian Thresholds: 'The Horrible (and Fortuitous) Inside-Outside That Real Space Is." *Perspecta* 19 (1982): 43–57.
———. *Ruskin's Maze: Mastery and Madness in His Art.* Princeton: Princeton University Press, 1981.
Fishman, Robert. *Urban Utopias in the Twentieth Century: Ebenezer Howard, Frank Lloyd Wright, Le Corbusier.* Cambridge: MIT Press, 1982.
Flinn, John J. *Official Guide to the World's Columbian Exposition.* Souvenir Edition. Chicago: Columbian Guide Company, 1893.
Fogarty, Frank. "Trailer Parks: The Wheeled Suburbs." *Architectural Forum* 111 (July 1959): 127–31.
Ford, Mark. *Raymond Roussel and the Republic of Dreams.* Ithaca, N.Y.: Cornell University Press, 2000.
Foucault, Michel. *Discipline and Punish.* Translated by Alan Sheridan. New York: Vintage, 1979.
———. *The Order of Things.* New York: Vintage, 1994.
———. "Other Spaces: The Principles of Heterotopia." *Lotus International,* nos. 48–49 (1986): 9–17.
———. *Power/Knowledge.* Edited by Colin Gordon. New York: Pantheon, 1980.
Frazer, Robert Walter. *Forts and Supplies: The Role of the Army in the Economy of the Southwest, 1846–1861.* Albuquerque: University of New Mexico Press, 1983.
Galveston Laboratory. "History of Chinese Shrimping." Galveston, Tex.: Southeast Fisheries Center. http://galveston.ssp.nmfs.gov/shrimpfishery/history.htm.
Gamble, Robert. "Some Recollections of the Seminole Chief Arpioka—Bowlegs—and His War with the States." Richard Keith Call Papers, Florida Historical Society, University of South Florida. N.p., n.d.
Geniné, Elizabeth, and William Geniné. *Camping with the Family.* Public Affairs Pamphlet No. 388. New York: Public Affairs Committee, 1966.
Glassie, Henry. *Material Culture.* Bloomington: Indiana University Press, 1999.

———. *Vernacular Architecture.* Bloomington: Indiana University Press, 2000.

Gorham, B. W. *Camp Meeting Manual: A Practical Book for the Camp Ground.* Boston: H. V. Degen, 1854.

Greene, David. "Gardener's Notebook." In *Archigram,* edited by Peter Cook, 110–15. New York: Princeton Architectural Press, 1999.

Gregotti, Vittorio. *Inside Architecture.* Translated by Peter Wong. Cambridge: MIT Press, 1996.

———. "Territory and Architecture." *Architectural Design Profile* 59, nos. 5–6 (1985): 28–34.

Groth, Paul. "Lot, Yard, and Garden: American Distinctions." *Landscape* 30, no. 3 (1990): 29–35.

Guthrie, John J. "Seeking the Sweet Spirit of Harmony: Establishing a Spiritualist Community at Cassadaga, Florida, 18933–1933." *Florida Historical Quarterly* 77, no. 1 (Summer 1998): 1–38.

Guthrie, John J., Phillip Charles Lucas, and Gary Monroe, eds. *Cassadaga: the South's Oldest Spiritualist Community.* Gainesville: University Press of Florida, 2000.

Hallock, Charles, ed. *Camp Life in Florida.* New York: Forest and Stream Publishing Co., 1876.

Hamlin, Talbot. *Benjamin Henry Latrobe.* New York: Oxford University Press, 1955.

Handke, Peter. *Across.* Translated by Alan Sheridan. New York: Norton, 1977.

Handy, M. P., ed. *Official Catalogue of Exhibits: World's Columbian Exhibition.* Chicago: W. B. Conkey, 1893.

Hann, John H., and Bonnie G. McEwan. *The Apalachee Indians and Mission San Luis.* Gainesville: University Press of Florida, 1998.

Harbison, Robert. *Eccentric Spaces.* New York: Avon, 1980.

Harrison, Jim. *Off to the Side.* New York: Grove, 2002.

Hart, Vaughan, and Peter Hicks, eds. *Paper Palaces: The Rise of the Renaissance Architectural Treatise.* New Haven: Yale University Press, 1998.

Hauptmann, Deborah. "The Past Which Is: The Present That Was." In *Cities in Transition,* edited by Arie Graafland and Deborah Hauptmann, 350–61. Rotterdam: 010 Publishers, 2001.

Hays, K. Michael. *Sanctuaries: The Last Works of John Hejduk.* New York: Whitney, 2002.

Heidegger, Martin. "Addendum to the 'Origin of the Work of Art.'" In *Poetry, Language, Thought,* by Heidegger, translated by Albert Hofstadter, 82–87. New York: Harper and Row, 1975.

———. "Building, Dwelling, Thinking." Translated by Albert Hofstadter. In *Basic Writings,* by Heidegger, 319–40. New York: Harper and Row, 1977.

Hejduk, John. *Riga.* Berlin: Aedes, 1988.

———. *Such Places as Memory.* Cambridge: MIT Press, 1998.

———. *Vladivostok.* New York: Rizzoli, 1989.

Henshall, James A. *Camping and cruising in Florida: An account of two winters passed in cruising around the coasts of Florida as viewed from the standpoint of an angler, a sportsman, a yachtsman, a naturalist and a physician.* Cincinnati: R. Clarke and Co., 1888. Gainesville: University of Florida, Florida Heritage Collection. http://fulltext.fcla.edu/cgi/t/text/text-idx?c=fhp&idno=FS00000044&format=pdf.

Hesselman, W. H. "Ode to the TCT (Tin Can Tourists)." Collection of Florida State Archives, Tallahassee.

Hibbard, Benjamin Horace. *A History of the Public Land Policies.* Madison: University of Wisconsin Press, 1965.

Hodes, Barnet, and Gale Roberson. *The Law of Mobile Homes.* 3rd ed. Washington, D.C.: Bureau of National Affairs, 1974.

Hofstadter, Albert. Introduction to *Poetry, Language, Thought,* by Martin Heidegger, ix–xxv. New York: Harper and Row, 1975.

Holl, Steven. *Rural and Urban House Types in North America.* New York: Pamphlet Architecture, 1982.

Homer. *Iliad.* Translated by A. T. Murray. Cambridge, Mass.: Loeb Classical Library, 1999.

hooks, bell. *Yearning: Race, Gender, and Cultural Politics.* Boston: South End Press, 1990.

Horn, Gillian. "Everyday in the Life of a Caravan." In *The Everyday and Architecture,* Architectural Design Profile No. 134, edited by Sarah Wigglesworth and Jeremy Till, 28–30. London: Architectural Design, 1998.

Huizinga, Johann. *Homo Ludens: A Study of the Play-Element in Culture.* Translated by R.F.C. Hull. Boston: Beacon Press, 1955.

Hulme, Keri. *Te Kaihau / The Windeater.* Wellington, New Zealand: Victoria University Press, 1986.

Hurston, Zora Neale. "Proposed Recording Expedition into the Floridas." Library of Congress, WPA Collections, 1939. http://memory.loc.gov/cgi-bin/query/r?ammem/flwpa:@field(DOCID+essay1).

Isherwood, Christopher. *World in the Evening.* New York: Random House, 1954.

Jackson, John Brinckerhoff. *Discovering the Vernacular Landscape.* New Haven: Yale University Press, 1984.

———. *The Necessity for Ruins.* Amherst: University of Massachusetts Press, 1980.

James, William. *The Meaning of Truth.* Cambridge: Harvard University Press, 1975.

———. *Pragmatism.* New York: Meridian Books, 1955.

Jameson, Fredric. *The Geopolitical Aesthetic.* Bloomington: Indiana University Press, 1995.

———. *The Seeds of Time.* New York: Columbia University Press, 1994.

Jencks, Charles, and Nathan Silver. *Ad Hocism: The Case for Improvisation.* New York: Doubleday, 1972.

Jessup, Elon. *The Motor Camping Book.* New York: Putnam's, 1921.

Kawamata, Tadashi. *Field Work.* Edited by Karin Orchard. Hannover: Sprengel Museum, 1998.

———. *Kawamata: Toronto Project 1989.* Toronto: Mercer Union, 1989.

Kellogg, Alice M. "Recent Camp Architecture, Part 1." *International Studio* 25 (March–June 1905): 73–76.

———. "Recent Camp Architecture, Part 2." *International Studio* 26 (July–October 1905): 6–10.

Kennedy, Stetson. *Palmetto Country.* Tallahassee: Florida A&M University Press, 1989.

Kimball, Winfield A., and Maurice H. Decker. *Touring with Tent and Trailer.* New York: Whittlesey House, 1936.

Kniffen, Fred. "The American Agricultural Fair: The Pattern." *Annals of the Association of American Geographers* 39, no. 4 (December 1949).

Knight, Diana. *Barthes and Utopia: Space, Travel, and Writing.* Oxford: Clarendon Press, 1997.

Kronenburg, Robert, ed. *Ephemeral/Portable Architecture.* Architectural Design Profile No. 135. London: Architectural Design, 1998.

———. *Houses in Motion: The Genesis, History and Development of the Portable Building.* 2nd ed. London: Wiley-Academic, 2002.

Kwinter, Sanford. *Architectures of Time: Toward a Theory of the Event in Modernist Culture.* Cambridge: MIT Press, 2001.

Kwon, Miwon. *One Place after Another: Site-Specific Art and Locational Identity.* Cambridge: MIT Press, 2002.

———. "The Wrong Place." *Art Journal* 59, no.1 (Spring 2000): 32–43.

Labov, William. *The Study of Nonstandard English.* Champaign, Ill.: National Council of Teachers of English, 1970.

Lavin, Sylvia. *Quatremère de Quincy and the Invention of Modern Language of Architecture.* Cambridge: MIT Press, 1992.

Lefebvre, Henri. *The Production of Space.* Translated by Donald Nicholson-Smith. Oxford: Blackwell, 1991.

———. *Writings on Cities.* Translated by Eleonore Kofman and Elizabeth Lebas. Oxford: Blackwell, 1996.

Leland, John. "Trying to Stay Put in Florida Mobile Homes." *New York Times*, June 22, 2003.

Lerup, Lars. *Building the Unfinished: Architecture and Human Action.* Beverly Hills, Calif.: Sage, 1977.

Levi-Strauss, Claude. *The Savage Mind.* Chicago: University of Chicago Press, 1966.

Liddell, H. G., ed. *Greek-English Lexicon.* Oxford: Oxford University Press, 1989.

Linn, Charles. "Field Station." *Architectural Record* (November 1992): 74–79.

Lippard, Lucy. *The Lure of the Local: Senses of Place in a Multicentered Society.* New York: New Press, 1997.

———. *On the Beaten Track: Tourism, Art, and Place.* New York: New Press, 1999.

Lockwood, C. C. *Atchafalaya: America's Largest River Basin Swamp.* Baton Rouge: Beauregard Press, 1981.

Louisiana Boundary Case. 394 U.S. 11, 32 (1969). In Aaron L. Shalowitz and Michael W. Reed, "The Tidelands Litigation," in *Shore and Sea Boundaries,* vol. 3, pt. 1, 92. Washington, D.C.: Office of Coast Survey, National Oceanic and Atmospheric Administration, 2000.

———. 394 U.S. 66 (1969). In Aaron L. Shalowitz and Michael W. Reed, "The Tidelands Litigation," in *Shore and Sea Boundaries,* vol. 3, pt. 1, 64. Washington, D.C.: Office of Coast Survey, National Oceanic and Atmospheric Administration, 2000.

Low, Thomas E. "The Chautauquans and Progressives in Florida." Edited by Beth Dunlop. *Journal of Decorative and Propaganda Arts, 1875–1945,* no. 23 (1998): 306–21.

MacCannell, Dean. *Empty Meeting Grounds: The Tourist Papers.* New York: Routledge, 1992.

Machiavelli, Niccolo. *The Art of War.* New York: Bobbs-Merrill, 1965.

Mallgrave, Harry Francis. Introduction to *The Four Elements of Architecture and Other Writings,* by Gottfried Semper. New York: Cambridge University Press, 1989.

Manatee County, Clerk of Circuit Court, Incorporation Book C: 335–7, Bradenton, Fla.

Marpillero, Sandro. "Strategic Thresholds." *Daidalos* 71 (1999): 48–61.

Maygarden, Benjamin D., and Jill-Karen Yakubik. *Bayou Chene: The Life Story of an Atchafalaya Basin Community.* New Orleans: U.S. Army Corps of Engineers, 1999. http://www.mvn.usace.army.mil/prj/hist_arch/bayou_chene/bayou_chene.pdf.

McCoy, Esther. *Vienna to Los Angeles: Two Journeys.* Santa Monica, Calif.: Arts + Architecture Press, 1979.

McDonough, William. "Selling Sarasota: Architecture and Propaganda in a 1920s Boom Town." *Journal of Decorative and Propaganda Arts, 1875–1945,* no. 23, edited by Beth Dunlop (1998): 10–31.

McEwan, Bonnie G., ed. *The Spanish Missions of La Florida.* Gainesville: University Press of Florida, 1993.

McEwen, Indra Kagis. *Socrates' Ancestor: An Essay in Architectural Beginnings.* Cambridge: MIT Press, 1993.

Miller, A. McA. "John Ruskin's American Utopias: A Survey of Ruskinite Intentional Communities, Colleges, and Universities in Tennessee, Georgia, Missouri, Illinois, and Florida (1894–1967)." Typescript in Miller's possession.

"Mobile Home Park Believed Oldest." *Sarasota Herald-Tribune,* November 15, 1964.

Mohney, David, and Keller Easterling, eds. *Seaside.* New York: Princeton Architectural Press, 1992.

Moore, Charles. "Trailers." In *Home Sweet Home: American Domestic Architecture,* edited by Charles Moore, Kathryn Smith, and Peter Becker, 49–51. New York: Rizzoli, 1983.

Moore, William D. "'To Hold Communion with Nature and the Spirit-World': New England's Spiritualist Camp Meetings, 1865–1910." In *Exploring Everyday Landscapes: Perspectives in Vernacular Architecture 7,* edited by Annmarie Adams and Sally McMurry, 230–50. Knoxville: University of Tennessee Press, 1997.

More, Thomas. *Utopia.* Translated by Robert M. Adams. Cambridge: Cambridge University Press, 1995.

Morley, David. *Home Territories: Media, Mobility and Identity.* New York: Routledge, 2000.

Morris, Dan, and Inez Morris. *The Weekend Camper.* New York: Bobbs-Merrill, 1973.

Morris, Robert. "Some Notes on the Phenomenology of Making." In *Continuous Project Altered Daily,* by Morris. Cambridge: MIT Press, 1995.

Mueller, Edward A. *Ocklawaha River Steamboats.* DeLeon Springs, Fla.: Painter, 1983.

Mugerauer, Robert. "Phenomenology and Vernacular Architecture." In *Encyclopedia of Vernacular Architecture of the World,* edited by Paul Oliver, 4 vols. Cambridge: Cambridge University Press, 1997.

Murphy, Richard J. *Authentic Visitors' Guide to the World's Columbian Exposition and Chicago,* May 1 to October 30, 1893. Chicago: Union News Company, 1893.

Nietzsche, Friedrich. *The Gay Science.* Translated by Josefine Nauckhoff. New York: Cambridge, 2001.

———. "Homeless" ("Ohne Heimat"). 1859. Translated by R. J. Hollingdale. In *Nietzsche: The Man and His Philosophy.* New York: Cambridge University Press, 1999.

———. *Human, All Too Human.* 2nd ed. Edited and translated by R. J. Hollingdale. New York: Cambridge University Press, 1996.

Nieuwenhuys, Constant. "New Babylon" (1974). In *Theory of the Dérive and Other Situationist Writings on the City,* edited by Libero Andreotti and Xavier Costa, 154–69. Barcelona: Museu d'Art Contemporani de Barcelona, 1996.

———. "New Urbanism." *Provo,* no. 9 (1966); English translation published by the

Friends of Malatesta, Buffalo, N.Y., 1970. Available online at: http://www.notbored.org/new-urbanism.html.

Nolen, John. *New Towns for Old.* Boston: Marshall Jones, 1927.

O'Grady, Patricia. "Thales of Miletus." In The Internet Encyclopedia of Philosophy. http://www.utm.edu/research/iep/t/thales.htm.

Original Constitution of the Tin Can Tourists. 1920. Collection of Florida State Archives, Tallahassee.

The Oxford English Dictionary. 2nd ed. 20 vols. Prepared by J. A. Simpson and E.S.C. Weiner. New York: Oxford University Press, 1989.

Padgett, H. R. "The Marine Shell Fisheries of Louisiana." Ph.D. diss., Louisiana State University, 1960.

Palmer, Robert E. A. *Studies of the Northern Campus Martius in Ancient Rome.* Philadelphia: American Philosophical Society, 1990.

Pearson, Jason, and Mark Robbins, eds. *University-Community Design Partnerships: Innovations in Practice.* New York: Princeton Architectural Press, 2002.

Peirce, Charles Sander. *The Collected Papers.* Edited by Charles Hartshorne and Paul Weiss. Cambridge: Belknap Press of Harvard University Press, 1960.

Pendleton-Jullian, Ann M. *The Road Is Not a Road and the Open City, Ritoque, Chile.* Cambridge: MIT Press, 1996.

Perez, Rodrigo. "Philippine Architecture." *AIA Journal* 34 (October 1960): 40–46.

———. "Tausug," In *Encyclopedia of Vernacular Architecture of the World,* edited by Paul Oliver, 4 vols. Cambridge: Cambridge University Press, 1997.

Pettena, Gianni, ed. *Superstudio, 1966–1982.* Florence: Electa, 1982.

Phelps, Dorothy Ann. "A Singular Land Use in the California Desert." Master's thesis, University of California at Riverside, 1989.

Pike, Sarah M. "Desert Goddesses and Apocalyptic Art: Making Sacred Spaces at the Burning Man Festival." In *God in Details: American Religion in Popular Culture,* edited by Eric Michael Mazur and Kate McCarthy, 155–76. New York: Routledge, 2001.

Plunket, Robert. "Walker Evans, the Mangrove Coast, and Me." In *Walker Evans: Florida,* by Walker Evans. Los Angeles: J. Paul Getty Museum, 2000.

Post, Emily. *By Motor to the Golden Gate.* New York: Appleton, 1917.

"Preliminary Problems in Constructing a Situation." Translated by Ken Knabb. Berkeley, Calif.: Bureau of Public Secrets, 1958. http://www.bopsecrets.org/SI/1.situations.htm.

Preston, Howard Lawrence. *Dirt Roads to Dixie: Accessibility and Modernization in the South, 1885–1935.* Knoxville: University of Tennessee Press, 1991.

Prizeman, Mark. "Intensity: Portable Architecture as Parable." In *Ephemeral/Portable Architecture,* Architectural Design Profile No. 135, edited by Robert Kronenburg, 22–29. London: Architectural Design, 1998.

Prochnik, George. *Putnam Camp: Sigmund Freud, James Jackson Putnam, and the Purpose of American Psychology.* New York: Other Press, 2006.

Quantrill, Malcolm, and Bruce Webb. Preface to *Urban Forms, Suburban Dreams,* edited by Quantrill and Webb. College Station: Texas A&M University Press, 1993.

Ray, Mary-Ann. "Gecekondu." In *Architecture of the Everyday,* edited by Steven Harris and Deborah Berke, 153–65. New York: Princeton Architectural Press, 1997.

Reeves, William Dale. *Westwego: From Cheniere to Canal.* Westwego, La.: Daniel P. Alario, 1996.

Reps, John. *Cities of the American West: A History of Frontier Urban Planning.* Princeton: Princeton University Press, 1980.

———. *Town Planning in Frontier America.* Princeton: Princeton University Press, 1969.

Ring, Alfred A. "The Mobile Home." *Urban Land* 25, no. 7 (July-August 1966): 1–6.

"Roadside Cabins for Tourists." *Architectural Record* 74 (December 1933): 457–62.

Robbins, H. E. "History of Braden Castle, Florida: A Soliloquy by the Old Castle in History and Romance—A True Story." Typescript in the collection of the Braden Castle Association, Bradenton, Manatee County, Fla., n.d., n.p.

Robinson, Lori, and Bill de Young. "Socialism in the Sunshine: The Roots of Ruskin, Florida." *Tampa Bay History* 4, no. 1 (Spring/Summer 1982): 5.

Robinson, W. W. *Land in California: The Story of Mission Lands, Ranchos, Squatters, Mining Claims, Railroad Grants, Land Scrip, Homesteads.* Berkeley and Los Angeles: University of California Press, 1948.

Roussel, Raymond. *How I Wrote Certain of My Books.* Edited and translated by Trevor Winkfield. New York: Sun, 1977.

Rudolph, Paul. Interview by Robert Bruegmann. February 28, 1986. In "Chicago Architects Oral History Project," Ernest R. Graham Study Center for Architectural Drawings, Department of Architecture, Art Institute of Chicago. Text online at: http://www.artic.edu/aic/collections/dept_architecture/rudolph.pdf.

Ruskin, John. *The Brantwood Diary of John Ruskin.* Edited by Helen Gill Viljoen. New Haven: Yale University Press, 1971.

———. *St. Mark's Rest.* London: George Allen, 1904.

———. *The Works of John Ruskin.* Edited by E. T. Cook and Alexander Wedderburn. Library Edition. London: George Allen, 1903–12.

Sadler, Simon. *The Situationist City.* Cambridge: MIT Press, 1999.

Sarnitz, August. *R. M. Schindler Architect: A Pupil of Otto Wagner between International Style and Space Architecture.* New York: Rizzoli, 1988.

Schama, Simon. *Landscape and Memory.* New York: Knopf, 1995.

Scheine, Judith. *R. M. Schindler.* New York: Phaidon, 2001.

Schuck, A. *Die Stabkarten der Marshall–Insulaner.* Hamburg: Persiehl, 1902.

Schulte, Karin. "From Nomad's Tent to Multimedia Vision: A Short History of Temporary Buildings." In *Temporary Buildings,* edited by Schulte, 20–50. New York: Gingko, 2000.

Segrest, Robert. "The Perimeter Projects: Notes for Design." *Assemblage* (1986): 24–35.

———. "The Perimeter Projects: The Architecture of the Excluded Middle." *Perspecta 23: Yale Architectural Journal* (1987): 54–65.

Semper, Gottfried. *The Four Elements of Architecture and Other Writings.* Translated by Harry Francis Mallgrave and Wolfgang Herrmann. New York: Cambridge University Press, 1989.

Serres, Michel. *Angels: A Modern Myth.* Translated by Frances Cowper. Paris: Flammarion, 1993.

———. *Genesis.* Translated by Geneviève James and James Nielson. Ann Arbor: University of Michigan Press, 1995.

———. *Hermes: Literature, Science, and Philosophy.* Edited by Josué V. Harari and David F. Bell. Baltimore: Johns Hopkins University Press, 1982.

———. *The Natural Contract.* Translated by Elizabeth MacArthur and William Paulson. Ann Arbor: University of Michigan Press, 1995.

———. *The Parasite.* Translated by Lawrence R. Schehr. Baltimore: Johns Hopkins University Press, 1982.

———. *The Troubadour of Knowledge.* Translated by Sheila Faria Glaser and William Paulson. Ann Arbor: University of Michigan Press, 1997.

Serres, Michel, and Bruno Latour. *Conversations on Science, Culture, and Time.* Translated by Roxanne Lapidus. Ann Arbor: University of Michigan Press, 1995.

Shalowitz, Aaron L., and Michael W. Reed. "United Nations Conferences on the Law of the Sea." In *Shore and Sea Boundaries,* vol. 1. Washington, D.C.: Office of Coast Survey, National Oceanic and Atmospheric Administration, 1962.

Shields, Rob. *Places on the Margin: Alternative Geographies of Modernity.* New York: Routledge, 1991.

Shrader, Welman A., ed. *Florida from the Air.* Vero Beach, Fla.: Aero-Graphic Corporation, 1936.

Siegal, Jennifer. *Mobile: The Art of Portable Architecture.* New York: Princeton Architectural Press, 2002.

Smith, Elizabeth A. T., and Michael Darling, eds. *The Architecture of R. M. Schindler.* Los Angeles: Museum of Contemporary Art, 2001.

Smith, McGregor W., Jr., *Thank You, Marco Polo: The Story of the First Around-the-World Trailer Caravan.* Coral Gables, Fla.: Wake-Brook House, 1966.

Sobieszek, Robert A. *Robert Smithson: Photo Works.* Los Angeles: Los Angeles County Museum of Art, 1993.

Solà-Morales, Ignasi, de. *Differences: Topographies of Contemporary Architecture.* Cambridge: MIT Press, 1997.

———."Mediations in Architecture and in the Urban Landscape," In *Cities in Transition,* edited by Arie Graafland and Deborah Hauptmann, 276–85. Rotterdam: 010 Publishers, 2001.

Somol, R. E. "Dummy Text, or the Diagrammatic Basis of Contemporary Architecture." In *Diagram Diaries,* by Peter Eisenman. New York: Universe, 1999.

Sontag, Susan. "Notes on Camp." In *Against Interpretation and Other Essays,* by Sontag, 275–92. New York: Dell, 1964.

Spigel, Lynn. "Media Homes: Then and Now." *International Journal of Cultural Studies* 4, no. 4 (2001): 385–411.

Stanaback, Richard. *A History of Hernando County.* Brooksville, Fla.: Hernando County Historical Society, 1976.

Stannard, Sandra J. "Trailers: Challenging a Tradition of Permanence and Place." *Traditional Dwellings and Settlements Review* 94 (1996): 55–70.

Sterling, Bruce. "Variation on a Theme Park (Taking the Kids to Burning Man)." In *Burning Man,* edited by Brad Wieners, n.p. San Francisco: Hardwired, 1997.

Stevens, Wallace. "Nomad Exquisite." In *The Palm at the End of the Mind,* edited by Holly Stevens, 44. New York: Vintage, 1990.

Stillman, William James. *Poetic Localities.* New York: Aperture, 1988.

Swanson, Betsy. *Historic Jefferson Parish from Shore to Shore.* New Orleans: Pelican, n.d.

Sykes, Mark. *The Caliph's Last Heritage: A Short History of the Turkish Empire.* London: Macmillan, 1915.

Taunton, Nina. "A Camp 'well planted': Encamped Bodies in 1590s Military Discourses and Chapman's *Caesar and Pompey.*" In *The Body in Late Medieval and Early Modern Culture,* edited by Darryll Grantley and Taunton, 83–96. Burlington, Vt.: Ashgate, 2000.

Thoreau, Henry David. *Walden.* Princeton: Princeton University Press, 2004.

Tin Can Tourists of the World. Unpublished documents, 1920–82. Florida State Archives, Tallahassee, Collection No. M93–2, boxes 2 and 3.

Truman, Ben C. *History of the World's Fair, Being a Complete Description of the World's Columbian Exposition from Its Inception.* Chicago: E. C. Morse and Co., 1893.

Tschumi, Bernard. *Architecture and Disjunction.* Cambridge: MIT Press, 1994.

Turner, Drexel. "Camptown Vitruvius: A Guide to the Gradual Understanding of Seaside, Florida." In *Urban Forms, Suburban Dreams,* edited by Malcolm Quantrill and Bruce Webb, 103–20. College Station: Texas A&M University Press, 1993.

Tze, Lao. *Tao Te Ching.* Translated by D. C. Lau. New York: Penguin, 1986.

Tzu, Sun. *The Art of War.* Edited by James Clavell. New York: Delacorte Press, 1983.

Ulmer, Gregory. "Abject Monumentality." *Lusitania* 1 (1993): 9–15.
———. *Heuretics.* Baltimore: Johns Hopkins University Press, 1994.
———. "Metaphoric Rocks: A Psychogeography of Tourism and Monumentality." In *The Florida Landscape: Revisited,* edited by Christoph Gerozissis, 39–50. Lakeland, Fla.: Polk Museum, 1992.
U.S. War Department. "Camps and Bivouac Area." In *Staff Officers' Field Manual: Organization, Technical and Logistical Data* (FM 101–10). Washington, D.C.: U.S. Government Printing Office, October 10, 1943.
———. "Shelters and Camps for Engineer Units." In *Engineer Field Manual: Operations of Engineer Field Units* (FM 5–6), 176–80 (pars. 156–63). Washington D.C.: U.S. Government Printing Office, April 23, 1943.
———. *Tents and Tent Pitching* (FM 20–15). Washington, D.C.: U.S. Government Printing Office, February 24, 1945, 51, 8–9. Updated on January 9, 1956.
Vallhonrat, Carles. "Tectonics Considered: Between the Presence and the Absence of Artifice." *Perspecta* (1988): 122–35.
van de Lippe, Klaar. Interview by Joep Van Lieshout. In *Ephemeral/Portable Architecture,* Architectural Design Profile No. 135, edited by Robert Kronenburg, 34–37. London: Architectural Design, 1998.
Vattimo, Gianni. *The End of Modernity.* Translated by Jon R. Snyder. Baltimore: Johns Hopkins University Press, 1988.
Vaughn, R. W. "Origin and History of the Camping Tourists of America." Braden Castle Association, dated January 8, 1927. Eaton Florida History Room, Manatee County Central Library, Bradenton, Fla.
Vico, Giambattista. *The New Science.* Ithaca, N.Y.: Cornell University Press, 1984.
Vidler, Anthony. *The Architectural Uncanny.* Cambridge: MIT Press, 1992.
———. "Diagrams of Utopia." *Daidalos* 74 (2000): 6–13.
Virilio, Paul. *The Aesthetics of Disappearance.* New York: Semiotext(e), 1991.
———. *Bunker Archeology.* New York: Princeton Architectural Press, 1994.
———. *A Landscape of Events.* Translated by Julie Rose. Cambridge: MIT Press, 2000.
Vitruvius. *Ten Books of Architecture.* Translated by Ingrid D. Rowland. Edited by Rowland and Thomas Noble Howe. Cambridge: Cambridge University Press, 2001.
Wallis, Allan D. *Wheel Estate: The Rise and Decline of Mobile Homes.* New York: Oxford University Press, 1991.
Webb, Michael, and David Greene. "Drive-in Housing." In *Archigram,* edited by Peter Cook. New York: Princeton Architectural Press, 1999.
Wehrly, Max S. "The Evolution of the House Trailer." *Urban Land* 26, no. 3 (March 1967).
Weiss, Ellen. *City in the Woods: The Life and Design of an American Camp Meeting on*

Martha's Vineyard. New York: Oxford University Press, 1987.

Weyrauch, Walter. *Gypsy Law: Romani Legal Traditions and Culture.* Berkeley and Los Angeles: University of California Press, 2001.

Wharton, Edward Ross. *Etyma Graeca: Etymological Lexicon of Classical Greek.* Chicago: Ares, 1974.

White, E. B. "On a Florida Key." In *Essays of E. B. White.* New York: Harper and Row, 1977.

Wieners, Brad, ed. *Burning Man.* San Francisco: Hardwired, 1997.

Wigley, Mark, and Catherine de Zegher, eds. *The Activist Drawing: Retracing Situationist Architectures from Constant's New Babylon and Beyond.* New York: Princeton Architectural Press, 2001.

Williams, S. Jeffress, Shea Penland, and Asbury H. Sallenger, eds. *Atlas of Shoreline Changes in Louisiana from 1853 to 1989.* Reston, Va.: U.S. Geological Survey and the Louisiana Geological Survey, 1992.

Willson, Corwin. "The Mobile House." *Architectural Record* (July 1936): 64–65.

Wing, Frank. "An Elegy in De Soto Park." *Tampa Morning Tribune,* April 4, 1924.

Wollen, Peter. "Bitter Victory: The Art and Politics of the Situationist International." In *On the Passage of a Few People through a Rather Brief Moment in Time: The Situationist International 1957–1972,* edited by Elisabeth Sussman, 20–61. Cambridge: MIT Press, 1989.

Wright, Frank Lloyd. *An Autobiography.* New York: Duell, Sloan and Pearce, 1943.

Wynne, Nick. *Tin Can Tourists of Florida 1900–1970.* Images of America series. Charleston, S.C.: Arcadia, 1999.

INDEX

Italicized page numbers refer to illustrations.